Springer Theses

Recognizing Outstanding Ph.D. Research

For further volumes:
http://www.springer.com/series/8790

Aims and Scope

The series "Springer Theses" brings together a selection of the very best Ph.D. theses from around the world and across the physical sciences. Nominated and endorsed by two recognized specialists, each published volume has been selected for its scientific excellence and the high impact of its contents for the pertinent field of research. For greater accessibility to non-specialists, the published versions include an extended introduction, as well as a foreword by the student's supervisor explaining the special relevance of the work for the field. As a whole, the series will provide a valuable resource both for newcomers to the research fields described, and for other scientists seeking detailed background information on special questions. Finally, it provides an accredited documentation of the valuable contributions made by today's younger generation of scientists.

Theses are accepted into the series by invited nomination only and must fulfill all of the following criteria

- They must be written in good English.
- The topic should fall within the confines of Chemistry, Physics, Earth Sciences, Engineering and related interdisciplinary fields such as Materials, Nanoscience, Chemical Engineering, Complex Systems and Biophysics.
- The work reported in the thesis must represent a significant scientific advance.
- If the thesis includes previously published material, permission to reproduce this must be gained from the respective copyright holder.
- They must have been examined and passed during the 12 months prior to nomination.
- Each thesis should include a foreword by the supervisor outlining the significance of its content.
- The theses should have a clearly defined structure including an introduction accessible to scientists not expert in that particular field.

James Keaveney

Collective Atom–Light Interactions in Dense Atomic Vapours

Doctoral Thesis accepted by
Durham University, UK

 Springer

Author
Dr. James Keaveney
Department of Physics
Durham University
Durham
UK

Supervisor
Prof. Charles Adams
Department of Physics
Durham University
Durham
UK

ISSN 2190-5053   ISSN 2190-5061   (electronic)
ISBN 978-3-319-38397-2   ISBN 978-3-319-07100-8   (eBook)
DOI 10.1007/978-3-319-07100-8
Springer Cham Heidelberg New York Dordrecht London

Printed on acid-free paper

Springer is part of Springer Science+Business Media (www.springer.com)

Supervisor's Foreword

The interaction of light and matter underlies many areas of science. Even though our understanding has advanced enormously there remain some fundamental questions that are very difficult to answer. For example, if one takes the simplest conceivable system of light-induced dipoles and then allows these dipoles to interact, the theoretical description quickly becomes unmanageable as we have $\mathcal{N}$ interacting particles, where $\mathcal{N}$ may be large, in an open quantum system. The interacting dipoles establish correlations that invalidate the mean field description of Maxwell's equations in a medium. However, if there is strong dephasing of the interacting dipoles the correlations are destroyed and a mean-field description is re-established. To understand these ideas requires detailed simple model systems where experimental data can be compared to theoretical simulations.

This thesis presents one approach to address such questions. The main results concern an experimental study of interacting atomic dipoles in a thin almost two-dimensional gaseous layer with a thickness in the range of 100 nm–1 μm. The layer consists of alkali atoms (with a single active valence electron) confined between two superpolished sapphire windows. The atoms are probed using a weak laser beam tuned through the main atomic resonance. The ability to compare the measured spectra with a model that provides an accurate description in the weakly interacting regime makes it possible to extract the interesting physics of strongly interacting dipoles. Due to the thermal motion of the atoms there is strong dephasing of the dipoles but, remarkably, it is still possible to observe coherent features of the dipole–dipole interaction. For example, as the length of the cell is varied one sees that the dipole–dipole-induced shift of the atomic resonance line oscillates with a period equal to one half the wavelength of the incident light. Historically, this propagation dependence of the mean-field shift is referred to as the collective or cooperative Lamb shift. This term arose because the dipole–dipole interactions within an ensemble may be modelled in terms of the exchange of virtual photons, similar to the self-interaction of an electron with a field of virtual photons. Despite the fact that these ideas have a long history, this thesis reports the first experimental measurement of the length dependence of the collective Lamb shift.

The experiment reported here has stimulated new theoretical work and further experiments that have enhanced our understanding of light–matter interactions in simple systems where the full many-body description is still challenging. This knowledge can now be exploited to design optical systems with the potential for new applications in quantum optics.

Durham, April 2014   Prof. Charles Adams

Abstract

This thesis presents an investigation of the fundamental interaction between light and matter, realised with a rubidium vapour confined in a cell whose thickness (in the propagation direction) is less than the optical wavelength. This confinement allows observation of spectroscopic features not found in longer cells, such as Dicke narrowing. These effects are measured experimentally and a theoretical model is developed, which allows the characterisation of the medium in terms of the atomic electric susceptibility. Interactions between the atoms and their surroundings, whether this be the walls of the cell or other nearby atoms, are explored. In the frequency domain we observe broadening and shifts of the spectral features due to these interactions. The atom–surface interaction shifts the spectral lines, following the expected $1/r^3$ van-der-Waals behaviour. The interatomic dipole–dipole interactions are more complex, and we find collective effects play an important role. We present an experimental verification of the full spatial dependence of the cooperative Lamb shift, which follows the theoretical prediction made 40 years ago, an important demonstration of coherent interactions in a thermal ensemble.

The interactions also play a role in determining the refractive index of the medium, limiting the maximum near-resonant index to $n = 1.31$. Using heterodyne interferometry, we experimentally measure an index of $n = 1.26 \pm 0.02$. This index enhancement leads to large bandwidth regions where a significant slow- or fast-light effect is present. We verify the fast-light effect in the time domain by observing the superluminal propagation of a sub-nanosecond optical pulse, and measure the group index of the medium to be $n_g = -1.0 \pm 0.1 \times 10^5$, the largest negative group index measured to date.

We investigate the radiative decay rate using time-domain fluorescence, and we observe radiation trapping effects in a millimetre-thickness vapour. Finally, we present results on sub-nanosecond coherent dynamics in the system which are achieved by pumping the medium with a strong optical pulse.

Acknowledgments

This thesis would not have been possible without the help and support of many people. First and foremost I would like to thank my supervisor, Charles Adams, without whom this thesis would not have been possible. His passion and enthusiasm for science is second to none, and his guidance and insight have been invaluable over the past few years. My thanks also go to Ifan Hughes for his valuable contributions to the project, innumerable enlightening conversations about physics and proof-reading of this thesis, as well as many 'fruitful discussions' about all things football. A special mention must go to our collaborators in Armenia, Armen and David, who make the finest, thinnest vapour cells in the world. Certainly without these cells this thesis would not be what it is today.

Thanks go to my office mates Lee (cheers, mush), Dan and Tim. Their friendship and banter has made the days spent in the office more enjoyable, particularly in the last few months whilst writing up. It is a great relief to know that I will never again be forced to listen to Radio 1 Xtra, courtesy of Chris. Questionable taste in music aside, it has been a pleasure to share a lab with him and I am fortunate to have benefitted from his expertise (and often his equipment) over the years. I am also grateful to Ulrich for teaching me the experimental skills and setting me off on the right track in the lab at the beginning of my Ph.D. and to Christophe and Rob for many discussions about dipole–dipole interactions. Kate joined the project in October 2012, and I take this opportunity to wish her luck for the future, as well as thank her for proof-reading a large part of this thesis.

I would like to extend my thanks to the rest of AtMol. I am sure that the friendly atmosphere and real camaraderie that exists in the group has a direct influence on its success, whether that be through stimulating discussions in the tea room or the willingness of every member to share their time and expertise to help someone else out.

My acknowledgments would not be complete without thanking the most important people in my life. To mam, dad, Veronica and Joseph—thank you for helping me achieve as much as I have done, for your constant encouragement in everything that I do. And finally, to Zara—the last three and a half years have had their ups and downs, and you've been there with me through all of it. Thank you for all of your love and support, I couldn't have done it without you.

Contents

Chapter 1
Introduction

Understanding the interaction between light and matter is of paramount importance to the scientific community across a wide range of disciplines. On the atomic scale, the light-matter interaction continues to be the focus of a large body of research. The award of the 2012 Nobel prize in physics to Serge Haroche and David Wineland[1] for their groundbreaking work with single atoms and ions in optical cavities stands as testament to this statement, with cavity quantum electrodynamics (QED) now a burgeoning area of research [1–3]. In cavity QED, the presence of the cavity enhances the atom-light interaction, leading to a modification of the intrinsic atomic properties such as its radiative lifetime.

1.1 Strong Interactions and Collective Effects

It is this enhancement of the atom-light interaction that this thesis deals with, though in a system without a cavity. Instead, the presence of another nearby atom can take the place of the cavity, modifying the interactions in a similar way.

The process responsible for this enhancement is the dipole–dipole interaction between two identical atoms, which causes the electric field between the atoms to be greatly enhanced when their separation is $<\lambda/2\pi$ (they are said to be in the near-field of each other), where λ is the transition wavelength between the two states [4, 5]. A similar effect has been previously observed using nanoplasmonics [6]. The strong interaction couples the two atoms together, so that their evolution is now fundamentally linked, which modifies the individual atomic energy levels and state lifetimes. As a consequence of these interactions, the behaviour of an ensemble of $\mathcal{N}$-atoms cannot simply be described by summing the response of a single atom $\mathcal{N}$ times. Instead, the presence of nearby atoms modifies the individual response.

The dipole-dipole interaction also has a coherent character, which has been studied extensively in ultracold atomic systems [7, 8], where motion is negligible and

---

[1] http://www.nobelprize.org/nobel_prizes/physics/laureates/2012

J. Keaveney, *Collective Atom–Light Interactions in Dense Atomic Vapours*,
Springer Theses, DOI: 10.1007/978-3-319-07100-8_1,
© Springer International Publishing Switzerland 2014

the coherence persists for an appreciable amount of time. In the literature these coherent processes have been known as either *collective* or *cooperative*. The difference between the two phenomena is subtle, but a recent paper by Javanainen et al. [9] has added some clarity to this debate. Essentially, they argue that in a homogeneously broadened ensemble (i.e. there is no motion), correlations between the dipoles are important and in this case we have *cooperative* interactions. These correlations can be thought of as a recurrent scattering of photons between atoms. In contrast to this, when there is inhomogeneous broadening, the recurrent scattering of photons is suppressed because each atom no longer has the same resonance frequency, and thus the correlations are suppressed. In this regime, the physics is that of a *collective* ensemble where a mean-field approximation is valid.

Both cooperative and collective interactions generated a great deal of interest in the community [10, 11], with much effort focussed on the phenomenon of superradiance [12, 13], the cooperative enhancement of the decay rate of an initially excited system. In recent years, there has been particular interest in using highly excited Rydberg systems to explore these interactions [7, 8], motivated in part by potential applications in quantum information systems [14]. There is also evidence of cooperative processes in natural systems, particularly in the light harvesting complexes responsible for photosynthesis [15, 16]. In these systems the interactions enable quantum transport of energy, vastly increasing the efficiency of the photosynthetic process [17, 18].

In this thesis, we investigate dipole-dipole interactions through spectroscopic measurements of a dense ensemble of thermal Rb atoms, extracting the properties of the interaction via a quantitative model of the atom-light interaction in the system.

The main challenge in observing these effects is that the dipoles need to be in the near-field of one another, separated by a distance $r < \lambda/2\pi$. The ground state transitions in Rb have near-infrared wavelengths with $\lambda \sim 800$ nm, so the interatomic spacing needs to be of the order of 100 nm. Assuming that the average separation between atoms $r_{\mathrm{av}} = 5/9\ N^{-1/3}$ (see Chap. 5), this means that an atomic density $N \sim 10^{14}$ cm^{-3} is required. For an ultracold atom experiment, this is difficult to achieve, requiring a Bose-Einstein condensate (BEC) to observe cooperative effects on an optical transition [19]. With thermal atoms, however, achieving these densities is trivial, as will be discussed in the next subsection.

As an alternative to increasing the density, one may instead exploit transitions which have longer wavelengths. Since the transition wavelengths between nearby states in highly-excited Rydberg atoms are much longer than for ground state atoms, placing two Rydberg atoms in the near field of each other requires much lower density. It is therefore not surprising that collective and cooperative interactions have been previously observed in these systems. Using electromagnetically induced transparency [20, 21], a cooperative atom-light interaction was demonstrated in an ultracold Rydberg gas [7], whilst recently a thermal Cs Rydberg experiment at moderate density demonstrated intrinsic optical bistability along with a superradiant cascade of excitation [22].

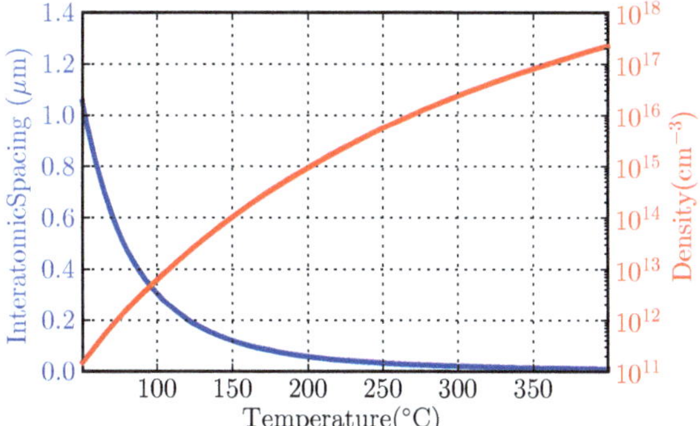

Fig. 1.1 Atomic number density N and average interatomic spacing $r_{av} = (5/9)N^{-1/3}$ as function of temperature

1.2 Thermal Vapours

In this thesis, a thermal (room temperature to $T \sim 350°$ C) atomic vapour is used to explore these strong interactions. We demonstrate collective interactions in low-lying excited states by raising the density of the atomic vapour to a point where the interactions dominate over every other dephasing mechanism in the system.

Thermal atomic vapours, typically confined in room temperature centimetre-sized glass cells, have been a workhorse in atomic physics over the years, used in a diverse range of experiments. Compared to a cold-atom setup, thermal experiments have the advantage of being relatively inexpensive and simple to set up. For the purposes of the experiments in this thesis, this remains a factor, but the main advantage of using a thermal vapour stems from the relationship between the temperature of a dilute vapour and the density of atoms.

The atomic number density, N, can be calculated from the vapour pressure p (the pressure of the gaseous phase in equilibrium with a solid or liquid bulk of the same material) and is given by [23]

$$\log_{10} p = 2.881 + 4.312 - \frac{4040}{T}$$
$$N = \frac{133.323 \times p}{k_\mathrm{B} T}, \tag{1.1}$$

for the liquid phase of Rb—the melting point of Rb (at zero pressure) is 312.46 K [24]. The inclusion or omission of the 2.881 term converts the pressure between Torr and atmospheres, respectively [23]. The factor 133.323 converts the vapour pressure from Torr to Pa. A pressure of 1 Torr corresponds to a number density $N \approx 2 \times 10^{16}$ cm^{-3}, which occurs at a temperature $T \approx 290°$ C. Equation (1.1) is perhaps the most important relationship for the experiments detailed in this thesis. The near-exponential scaling with temperature, shown in Fig. 1.1, allows for a vast parameter space that is continuously tunable simply via adjustment of the temperature. In experiments with ultracold atoms, it is possible to achieve densities of around 10^{11}–10^{12} cm^{-3} in a magneto-optical trap, and in a BEC densities of 10^{13}–10^{15} cm^{-3} are reasonable [25]. Thermal vapours can have densities orders of magnitude larger than even BECs, and tunable over a much wider range. The interactions between atoms considered in this work are resonant dipole-dipole interactions, scaling as $r^{-3} \propto N$, which by implication means we also have control over the average interaction strength over many orders of magnitude. It is possible to move smoothly from a regime where interactions are completely negligible to one where the interactions are the dominant effect, as we shall demonstrate in Chap. 5.

The disadvantage of thermal vapours comes from the atomic motion. This introduces unwanted Doppler broadening of the absorption lines from the natural linewidth (for the states this work is concerned with this is around 6 MHz) to around 500 MHz at room temperature. In addition, since Doppler broadening is inhomogeneous, each atom sees a different effect and so any coherent effect is usually washed out by a motional dephasing of the medium.

1.3 Nano-Cells

Increasing the density of the vapour can only go so far, until the optical depth of the medium becomes a limiting factor. Ignoring interactions between atoms for now (which reduce the effective optical depth), the optical depth is approximately proportional to the product of density and thickness ℓ of the medium. As one is increased the other must decrease by an according amount to maintain an equivalent optical depth. As an order-of-magnitude approximation, a room temperature 5 cm Rb vapour cell will absorb roughly half of the light when it is resonant with the ground state to first excited state transition. To realise the same optical depth at $T = 300°$ C, the thickness of the vapour must be reduced by 6 orders of magnitude, to just 50 nm. Other approaches to investigate high density vapours have been implemented in the past, such as selective reflection spectroscopy [26, 27] or using off-resonance excitation [28]. In all of these techniques the experiments can be kept as simple as possible; most involve only a single excitation laser and detection of transmitted or reflected light. Using a nano-cell allows access to the whole spectrum, instead of just the off-resonant wings where the optical depth is lower, and for the investigation of effects such as Dicke narrowing [29, 30], first observed in the microwave regime but extended to optical wavelengths with the use of sub-wavelength nano-cells [31]. Chapter 3 will discuss these effects in detail.

Creating a vapour cell with sub-wavelength thickness is extremely challenging, but the group of David Sarkisyan in Armenia has been successfully producing these sub-wavelength thickness 'nano-cells' for some years now. A photograph of the window region of the vapour cell employed in this work is shown in Fig. 1.2. This chapter is not concerned with the technical details (these are discussed in Appendix A), but here we summarise the important points. Clearly visible on the photograph are high

Fig. 1.2 Photograph of the nano-cell used for most of the work presented in this thesis, showing the Newtons rings interference pattern. At the *centre* of the rings, the gap between the windows, and hence the local vapour thickness, is just 30 nm, extending to 2 μm at the *bottom* of the windows

quality Newton's rings, an interference pattern that occurs because one of the two windows is slightly curved, causing the gap between the two windows to vary. For this particular cell, at the centre of the Newtons rings, the gap, and hence the vapour thickness ℓ is just 30 nm. The thickness extends outwards from here, and the maximum thickness is around 2 μm at the bottom of the windows. We can therefore scale the local vapour thickness over nearly two orders of magnitude, which combined with the tunable density provides us with a very large parameter space to investigate.

Recently there has been a resurgence of interest in thermal vapours confined in thin cells. As the total size of these thin cells can be reduced down to the milimetre scale [32, 33], there has been a great deal of interest in harnessing their potential as miniaturised atomic clocks [34, 35] and very sensitive magnetometers [36–38]. Nano-cells could also be useful as magnetometers, enabling sub-micron resolution of magnetic fields in one dimension [39]. Making further use of the atomic sensitivity to magnetic fields, by placing the cells in a high field found inside a strong permanent ring magnet, it has been possible to realise a narrowband optical isolator [40].

Exciting Rydberg atoms in these miniature cells opens the door to applications in quantum information processing [14]. Arrays of micro-cells have been proposed as a scalable architecture for constructing Rydberg atom quantum gates [41], and the same technology has been proposed as a single photon source [42].

1.4 Structure of this Thesis

The remainder of this thesis is organised as follows:

- In Chap. 2 we consider a simple model for the atom-light interaction, neglecting any interaction between the atoms, which forms the basis of the models used throughout this thesis.

- Chapter 3 builds on the results from the previous chapter, and deals with the specific effects that are encountered with nano-cells such as Dicke narrowing of the spectral lines.
- Chapter 4 deals with the van der Waals atom-surface interaction in the system, an additional consequence of using such a short cell.
- In Chap. 5 we explore the dipole-dipole interaction between two identical atoms, which leads to resonant shifts and broadening of spectral features. We find dipole-dipole interactions are important in the medium, and experimentally verify the spatial oscillations of the Cooperative Lamb shift.
- Chapter 6 examines the real part of the susceptibility, which is responsible for the refractive index of the medium. We show that a Rb vapour can have a near-resonant refractive index comparable to that of water under the ideal conditions.
- Chapter 7 deals with the fast and slow light effects in the medium, a consequence of the large refractive index gradient in the medium. We demonstrate the largest negative group index that has been measured to date, which leads to a significant fractional advance (the so-called 'superluminal' propagation) of a sub-nanosecond optical pulse.
- In Chap. 8 we look at the time-resolved resonant fluorescence of the medium, after excitation with a sub-nanosecond optical pulse.
- Chapter 9 presents initial work towards understanding the coherent nature of the medium when excited by a strong optical pulse with sub-nanosecond duration.
- Finally, Chap. 10 presents an outlook for the project.

References

1. R. Miller et al., Trapped atoms in cavity QED: coupling quantized light and matter. J. Phys. B **38**, S551 (2005)
2. H. Walther, B.T.H. Varcoe, B.-G. Englert, T. Becker, Cavity quantum electrodynamics. Rep. Prog. Phys. **69**, 1325 (2006)
3. S. Haroche, J.-M. Raimond, *Exploring the Quantum: Atoms, Cavities, and Photons* (OUP, Oxford, 2006)
4. R. J. Bettles and C. S. Adams, Private, Communication, 2013.
5. L. Chomaz, L. Corman, T. Yefsah, R. Desbuquois, J. Dalibard, Absorption imaging of a quasi-two-dimensional gas: a multiple scattering analysis. New J. Phys. **14**, 055001 (2012)
6. S. Kühn, U. Håkanson, L. Rogobete, V. Sandoghdar, Enhancement of Single-Molecule Fluorescence Using a Gold Nanoparticle as an Optical Nanoantenna. Phys. Rev. Lett. **97**, 017402 (2006)
7. J.D. Pritchard et al., Cooperative Atom-Light Interaction in a Blockaded Rydberg Ensemble. Phys. Rev. Lett. **105**, 193603 (2010)
8. Y.O. Dudin, A. Kuzmich, Strongly interacting Rydberg excitations of a cold atomic gas. Science **336**, 887 (2012)
9. J. Javanainen, J. Ruostekoski, Y. Li, S.-M. Yoo, Shifts of a Resonance Line in a Dense Atomic Sample. Physical Review Letters **112**, 113603 (2014)
10. M. Gross, S. Haroche, Superradiance: An essay on the theory of collective spontaneous emission. Phys. Rep. **93**, 301 (1982)
11. M.O. Scully, A.A. Svidzinsky, The super of superradiance. Science **325**, 1510 (2009)

12. R. Dicke, Coherence in Spontaneous Radiation Processes. Phys. Rev. **93**, 99 (1954)
13. C. Hettich et al., Nanometer resolution and coherent optical dipole coupling of two individual molecules. Science **298**, 385 (2002)
14. M. Saffman, T. Walker, and K. Mølmer, Quantum information with Rydberg atoms, Rev. Mod. Phys. 82, 2313 (2010).
15. R. Monshouwer, M. Abrahamsson, F. van Mourik, R. van Grondelle, Superradiance and Exciton Delocalization in Bacterial Photosynthetic Light-Harvesting Systems. J. Phys. Chem. B **101**, 7241 (1997)
16. J.L. Herek, W. Wohlleben, R.J. Cogdell, D. Zeidler, M. Motzkus, Quantum control of energy flow in light harvesting. Nature **417**, 533 (2002)
17. G.S. Engel et al., Evidence for wavelike energy transfer through quantum coherence in photosynthetic systems. Nature **446**, 782 (2007)
18. G. Panitchayangkoon et al., Direct evidence of quantum transport in photosynthetic light-harvesting complexes., P. Natl. Acad. Sci. 108, 20908 (2011).
19. S. Inouye, Superradiant Rayleigh Scattering from a Bose-Einstein Condensate. Science **285**, 571 (1999)
20. S.E. Harris, J.E. Field, A. Imamoğlu, Nonlinear optical processes using electromagnetically induced transparency. Phys. Rev. Lett. **64**, 1107 (1990)
21. K. J. Boller, A. Imamoğlu, and S. E. Harris, Observation of Electromagnetically Induced Transparency, Phys. Rev. Lett. 66 (1991).
22. C. Carr, R. Ritter, K. J. Weatherill, and C. S. Adams, Cooperative non-equilibrium phase transition in a dilute thermal atomic gas, arxiv.org, 1302.6621 (2013), arXiv:1302.6621v1.
23. D. A. Steck, Rubidium 85 D Line Data, (2009).
24. D.K. Belashchenko, Molecular-dynamics simulation of the high-pressure properties of rubidium. High Temperature **48**, 646 (2010)
25. C.J. Pethick, H. Smith, *Bose-Einstein condensation in dilute gases* (CUP, Cambridge, 2001)
26. J. Woerdman, M.F.H. Schuurmans, Spectral narrowing of selective reflection from sodium vapour. Opt. Commun. **14**, 248 (1975)
27. M.F.H. Schuurmans, Spectral narrowing of selective reflection. J. Phys. (Paris) **37**, 469 (1976)
28. P. Siddons, C.S. Adams, I.G. Hughes, Off-resonance absorption and dispersion in vapours of hot alkali-metal atoms. J. Phys. B **42**, 175004 (2009)
29. R. Dicke, The Effect of Collisions upon the Doppler Width of Spectral Lines. Phys. Rev. **89**, 472 (1953)
30. R. Romer, R. Dicke, New Technique for High-Resolution Microwave Spectroscopy. Phys. Rev. **99**, 532 (1955)
31. S. Briaudeau, S. Saltiel, G. Nienhuis, D. Bloch, M. Ducloy, Coherent Doppler narrowing in a thin vapor cell: Observation of the Dicke regime in the optical domain. Phys. Rev. A **57**, R3169 (1998)
32. S. Knappe, V. Velichansky, H.G. Robinson, J. Kitching, L. Hollberg, Compact atomic vapor cells fabricated by laser-induced heating of hollow-core glass fibers. Rev. Sci. Instr. **74**, 3142 (2003)
33. L.-A. Liew et al., Microfabricated alkali atom vapor cells. Appl. Phys. Lett. **84**, 2694 (2004)
34. S. Knappe et al., A microfabricated atomic clock. Appl. Phys. Lett. **85**, 1460 (2004)
35. M. Pellaton, C. Affolderbach, Y. Pétremand, N. de Rooij, G. Mileti, Study of laser-pumped double-resonance clock signals using a microfabricated cell. Phys. Scr. **T149**, 014013 (2012)
36. W.C. Griffith, R. Jimenez-Martinez, V. Shah, S. Knappe, J. Kitching, Miniature atomic magnetometer integrated with flux concentrators. Appl. Phys. Lett. **94**, 023502 (2009)
37. S. Li, P. Vachaspati, D. Sheng, N. Dural, M.V. Romalis, Optical rotation in excess of 100 rad generated by Rb vapor in a multipass cell. Phys. Rev. A **84**, 061403 (2011)
38. D. Sheng, S. Li, N. Dural, M.V. Romalis, Subfemtotesla Scalar Atomic Magnetometry Using Multipass Cells. Phys. Rev. Lett. **110**, 160802 (2013)
39. A. Sargsyan et al., High contrast D1 line electromagnetically induced transparency in nanometric-thin rubidium vapor cell. Appl. Phys. B **105**, 767 (2011)

40. L. Weller et al., Optical isolator using an atomic vapor in the hyperfine Paschen-Back regime. Opt. Lett. **37**, 3405 (2012)
41. H. Kübler, J.P. Shaffer, T. Baluktsian, R. Löw, T. Pfau, Coherent excitation of Rydberg atoms in micrometre-sized atomic vapour cells. Nature Photon. **4**, 112 (2010)
42. M. M. Muller et al., Room temperature Rydberg Single Photon Source, arXiv.org, 1212.2811 (2012), arXiv:1212.2811v1.

Publications Arising from this Work

43. J. Keaveney et al., Cooperative Lamb Shift in an Atomic Vapor Layer of Nanometer Thickness. Phys. Rev. Lett. **108**, 173601 (2012)
44. J. Keaveney, I.G. Hughes, A. Sargsyan, D. Sarkisyan, C.S. Adams, Maximal Refraction and Superluminal Propagation in a Gaseous Nanolayer. Phys. Rev. Lett. **109**, 233001 (2012)
45. J. Keaveney et al., Optical transmission through a dipolar layer, arxiv.org, 1109.3669v2 (2011), arXiv:1109.3669v2.

Related Publications

46. A. Sargsyan, D. Sarkisyan, U. Krohn, J. Keaveney, C.S. Adams, Effect of buffer gas on electromagnetically induced transparency in a ladder system using thermal rubidium vapor. Phys. Rev. A **82**, 045806 (2010)

Chapter 2
Atom–Light Interactions for Independent Atoms

The purpose of this chapter is to introduce from a fundamental level the model for atom-light interactions that is used in much of the rest of this work. We will introduce a fully quantitative model of the atom-light interaction in thermal alkali-metal vapour that can be used to make predictions for and be compared to experimental spectroscopic data.

We will start our discussion with the most basic of systems, that of an oscillator with two levels coupled by a near-resonant driving field, before adding in the full atomic structure and motional effects.

The development of this model in the atomic and molecular physics group at Durham over recent years has led to a plethora of research [1–7], and this chapter summarises some of that work. In the next chapter we will expand this model to cover the additional complexities of spectroscopy in nano-cells, and in Chap. 5 we will add the effects of the dipole–dipole interaction. It is therefore important that the underlying model is well understood.

2.1 The Two-Level Atom

We start our discussion of atom-light systems by looking at the case where the atom has only two levels, coupled by a near-resonant monochromatic light source. Figure 2.1 shows a typical level scheme for the two level system.

We assume for simplicity in this picture that there is no decay out of the ground state. The ground and excited states are coupled by a near-resonant laser with Rabi frequency Ω. The excited state spontaneously decays, at a characteristic rate Γ_0, and the atom can only return to the ground state. For the first excited states of Rb, $\Gamma_0 \approx 2\pi \times 6\,\text{MHz}$. The detuning Δ is the difference in angular frequencies between the laser (ω_p) and atomic resonance (ω_0) frequencies, $\Delta = \omega_\text{p} - \omega_0$.

J. Keaveney, *Collective Atom–Light Interactions in Dense Atomic Vapours*,
Springer Theses, DOI: 10.1007/978-3-319-07100-8_2,

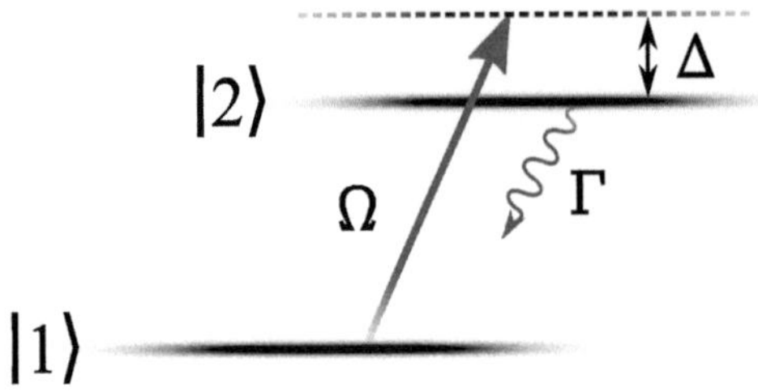

Fig. 2.1 Schematic of the 2-level atom. Ground- ($|1\rangle$) and excited-states ($|2\rangle$) are coupled by a near-resonant laser with angular frequency ω_p, driving Rabi frequency Ω, and detuned by $\Delta = \omega_p - \omega_0$ from resonance. The excited state spontaneously decays at a rate Γ_0

A full many-body quantum mechanical treatment of this system is intractable due to the large numbers involved[1] so we must employ some approximations. The first is known as the semiclassical approximation, where a quantised atomic system is coupled to a classical electromagnetic field, valid for large photon numbers. The atom-light interaction is then between the applied field and the electric dipole moment of the atom and we neglect higher order multipole terms—this is known as the electric-dipole approximation.

Since the atom number is intractably large, we need a statistical method of describing the properties of the system. The density matrix formalism provides such a description. As the orthonormal basis states $\{|i\rangle\}$ form a complete set, we can describe an individual quantum state of the system as a superposition of basis states

$$|\psi^s\rangle = \sum_i C_i^s |i\rangle, \tag{2.1}$$

where the C_i^s coefficients are probability amplitudes of being in the basis state $|i\rangle$. When the system is in a mixed state it is not possible to describe it in terms of a ket vector. In this case we can use the density matrix, defined as

$$\rho = \sum_s p(s)|\psi^s\rangle\langle\psi^s|, \tag{2.2}$$

where $p(s)$ is the probability of the pure state s. For an n-level system, the density matrix is an n-by-n matrix where the diagonal terms ρ_{nn} represent the populations of each level and the off-diagonal terms ρ_{mn} correspond to the coherences between levels.

One can calculate the expectation value of the Hamiltonian (or any other operator) from

---

[1] For typical experimental parameters, we have 10^{22} atoms/m^3, a focal spot size $\sim 10\,\mu$m and a cell thickness of 0.5 μm, which gives roughly 10^6 atoms. 1 nW of incident laser power corresponds to roughly 10^9 photons per second and an intensity of 1 mW/cm^2.

$$\langle \hat{\mathcal{H}} \rangle = \mathrm{Tr}(\rho \hat{\mathcal{H}}), \tag{2.3}$$

where the trace is defined in the usual way as $\mathrm{Tr}(\hat{A}) = \sum_i A_{ii}$.

2.1.1 Hamiltonian

We want to calculate the time evolution of the atom-light system, so the problem we need to solve is a Schrödinger-type equation where the Hamiltonian of the combined atom-light system is given by the sum of the atomic Hamiltonian $\hat{\mathcal{H}}^{\mathrm{a}}$ and the Hamiltonian due to the interaction with the light field $\hat{\mathcal{H}}^{\mathrm{i}}$

$$\hat{\mathcal{H}} = \hat{\mathcal{H}}^{\mathrm{a}} + \hat{\mathcal{H}}^{\mathrm{i}}. \tag{2.4}$$

We can describe the system in terms of the orthogonal basis states that make up the bare (unperturbed) atomic states (i.e. solutions to the time-independent Schrödinger equation $\hat{\mathcal{H}}^{\mathrm{a}}|i\rangle = E_i|i\rangle$). The Hamiltonian can be expressed in matrix form, with elements

$$\mathcal{H}_{ij} = \langle i|\hat{\mathcal{H}}|j\rangle. \tag{2.5}$$

The atomic part of the Hamiltonian contains information about the energies of the states. Setting the ground state energy as the zero of the energy scale, we have

$$\hat{\mathcal{H}}^{\mathrm{a}} = \begin{pmatrix} 0 & 0 \\ 0 & \hbar\omega_0 \end{pmatrix}. \tag{2.6}$$

The interaction Hamiltonian in the dipole approximation is given as

$$\hat{\mathcal{H}}^{\mathrm{i}} = -\hat{\boldsymbol{\mu}} \cdot \mathcal{E}, \tag{2.7}$$

where $\hat{\boldsymbol{\mu}} = e\,\hat{\mathbf{r}}$ is the dipole operator, which has only only off-diagonal elements (the diagonal terms $\langle 1|\hat{\boldsymbol{\mu}}|1\rangle = \langle 2|\hat{\boldsymbol{\mu}}|2\rangle = 0$ due to the parity of the states[2]). For monochromatic light the electromagnetic field, $\mathcal{E}$, can be expressed as

$$\mathcal{E} = \hat{\mathbf{e}}\mathcal{E}_0 \cos(\omega_p t) = \frac{\hat{\mathbf{e}}\mathcal{E}_0}{2}(e^{i\omega_p t} + e^{-i\omega_p t}), \tag{2.8}$$

where $\hat{\mathbf{e}}$ is a vector describing the polarisation of the field. Assuming the dipole moment is aligned with the field, the components of the interaction Hamiltonian can then be expressed as

---

[2] In the wavefunction representation, the matrix element between an initial and final state is given by $\langle f|\hat{\boldsymbol{\mu}}|i\rangle = e\int_{-\infty}^{+\infty} \psi_f^*(\mathbf{r})\,\mathbf{r}\,\psi_i(\mathbf{r})\mathrm{d}\mathbf{r}$. The dipole operator is an odd function, therefore for this integral to be non-zero, the two states need to be of opposite parity.

$$\mathcal{H}_{21}^i = \mathcal{H}_{12}^{i*} = -\frac{d_{21}\mathcal{E}_0}{2}(e^{-i\omega_p t} + e^{i\omega_p t}). \tag{2.9}$$

The first term in Eq. (2.9) represents the interaction when either the system absorbs a photon and is raised to the excited state, or when the system emits a photon and falls to the ground state from the excited state. The second term is known as the counter-rotating term, associated with the simultaneous absorption of a photon and falling from the excited to ground state, or when the system emits a photon and is raised to the excited state. These second terms are omitted from the rest of the analysis, an approximation known as the Rotating Wave Approximation (RWA) [8]. When the problem is transformed into a rotating frame, we see that the omitted terms are those owing to the 'fast' oscillations, which evolve as $e^{i(\omega_0 + \omega_p)t}$, and average away to zero. We are left with the 'slow' oscillations, which evolve as $e^{i(\omega_0 - \omega_p)t} = e^{-i\Delta t}$, which do not average away.

The total Hamiltonian can therefore be rewritten in the RWA as

$$\hat{H} = \hbar \begin{pmatrix} 0 & \dfrac{\Omega}{2}e^{i\omega_p t} \\ \dfrac{\Omega}{2}e^{-i\omega_p t} & \omega_0 \end{pmatrix}. \tag{2.10}$$

where the Rabi frequency Ω is given by

$$\Omega = -\frac{\boldsymbol{\mu}_{21} \cdot \hat{\mathbf{e}}\mathcal{E}_0}{\hbar} = -\frac{e\mathcal{E}_0}{\hbar}\langle 1|\mathbf{r}|2\rangle, \tag{2.11}$$

and we have assumed that the polarisation of the driving field is aligned with the atomic dipole moment.

2.1.2 Time Evolution

The time evolution of the system can then be calculated from the time dependent Schrödinger equation, which for the density matrix is equivalent to solving

$$\frac{d}{dt}\rho = \frac{1}{i\hbar}\left[\hat{H}, \rho\right], \tag{2.12}$$

the Liouville-von Neumann equation. However, this does not account for decay in the system due to, for example, spontaneous emission. Decay is added in phenomenologically as a statistical process, grouped into the decay matrix $\mathcal{L}$. The equation that must be solved is known as the Linblad master equation, and is given by

$$\frac{d}{dt}\rho = \frac{1}{i\hbar}\left[\hat{H}, \rho\right] - \mathcal{L}. \tag{2.13}$$

For the two level system, the steady state solutions can be derived analytically, which we do by writing (2.13) as a set of coupled rate equations known as the Optical Bloch Equations (OBE). For the two level system with dephasing due only to spontaneous decay, these are

$$\frac{\mathrm{d}\rho_{11}}{\mathrm{d}t} = -\frac{\mathrm{d}\rho_{22}}{\mathrm{d}t}$$

$$= \Gamma_0 \rho_{22} - \frac{\mathrm{i}\Omega}{2}(\tilde{\rho}_{21} - \tilde{\rho}_{12}) \tag{2.14}$$

$$\frac{\mathrm{d}\tilde{\rho}_{21}}{\mathrm{d}t} = \frac{\mathrm{d}\tilde{\rho}_{12}^{*}}{\mathrm{d}t}$$

$$= -(\Gamma_0/2 - \mathrm{i}\Delta)\tilde{\rho}_{21} - \frac{\mathrm{i}\Omega}{2}(\rho_{11} - \rho_{22}), \tag{2.15}$$

where we have introduced the 'slow' variables in the coherence terms, $\tilde{\rho}_{21} = \tilde{\rho}_{12}^{*} = \rho_{21}e^{-\mathrm{i}\omega_p t}$. To conserve population, we must have $\rho_{11} + \rho_{22} = 1$, and clearly $\mathrm{d}\rho_{11}/\mathrm{d}t + \mathrm{d}\rho_{22}/\mathrm{d}t = 0$. In the steady state, $\mathrm{d}\rho/\mathrm{d}t = 0$, and we thus find

$$\tilde{\rho}_{21} = -\frac{\mathrm{i}\Omega/2}{\Gamma_0/2 - \mathrm{i}\Delta}(\rho_{11} - \rho_{22}). \tag{2.16}$$

2.2 Electric Susceptibility

We can describe the optical response of the atomic system to an applied light field in terms of the complex electric susceptibility, χ, which describes how the medium polarises in response to an applied electric field. The macroscopic polarisation of the medium is related to the individual electric dipole moments of the atoms. For an ensemble we take the average dipole moment and obtain

$$\boldsymbol{P} = N\langle\hat{\boldsymbol{\mu}}\rangle = N\boldsymbol{\mu}_{21}[\rho_{21}e^{-\mathrm{i}\omega_p t} + \rho_{12}e^{\mathrm{i}\omega_p t}], \tag{2.17}$$

where N is the atomic density. The polarisation is related to the complex linear susceptibility by [9]

$$\boldsymbol{P} = (\epsilon - \epsilon_0)\mathcal{E} = \epsilon_0\chi\mathcal{E} \tag{2.18}$$

$$= \frac{\epsilon_0\mathcal{E}_0}{2}[\chi e^{-\mathrm{i}\omega_p t} + \chi^{*}e^{\mathrm{i}\omega_p t}], \tag{2.19}$$

assuming $\mathcal{E}_0$ is real. We combine Eqs. (2.17) and (2.19) and take the dot product with $\boldsymbol{\mu}_{21}$, yielding

Fig. 2.2 Real (*blue*) and imaginary (*red*) parts of the linear susceptibility for a two-level atom around a resonance

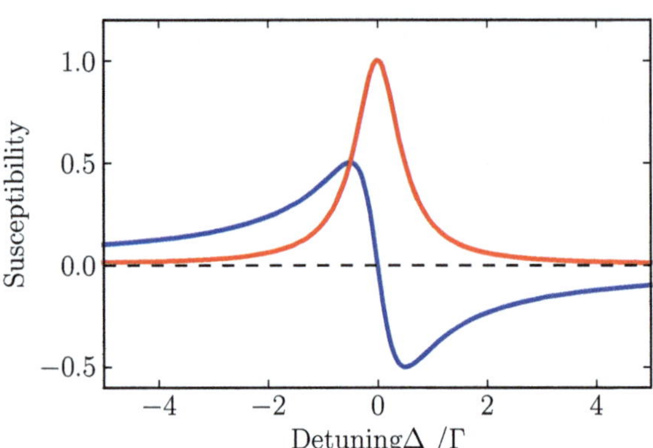

$$N\mu_{21}^2(\tilde{\rho}_{21}e^{-i\omega_p t} + \tilde{\rho}_{12}e^{i\omega_p t}) = -\frac{1}{2}\epsilon_0\hbar\Omega(\chi e^{-i\omega_p t} + \chi^* e^{i\omega_p t}). \tag{2.20}$$

We then find a relationship between the susceptibility and the coherence term in the density matrix

$$\chi = -\frac{2N\mu_{21}^2}{\epsilon_0\hbar\Omega}\tilde{\rho}_{21}. \tag{2.21}$$

Combining this with Eq. (2.16), in the weak probe limit, where $\Omega \ll \Gamma_0$ and we assume all the population remains in the ground state ($\rho_{11} - \rho_{22} \approx 1$), we obtain as a function of the laser detuning

$$\chi(\Delta) \simeq \frac{iN\mu_{21}^2}{\epsilon_0\hbar(\Gamma_0/2 - i\Delta)}. \tag{2.22}$$

In the absence of any other broadening mechanism, this has real and imaginary components with the classic dispersive and Lorentzian lineshapes, as shown in Fig. 2.2. The refractive index of the medium is related to its relative dielectric constant ϵ_r and thus the susceptibility of the medium, $n = \sqrt{\epsilon_r} = \sqrt{1 + \chi}$.

It should be noted that since much of this work deals with large susceptibilities, the usual approximation of $n \approx 1 + \chi/2$ does not hold. Once the susceptibility becomes large, local field effects need to be considered (discussed in Chap. 5). If one does not consider local field effects, $\sqrt{1 + \chi(\Delta)}$ is no longer symmetric about $\Delta = 0$ and one might mistakenly identify a fictitious blue-shift of the lines (see Appendix D).

A monochromatic electric field propagating through a medium with a (isotropic) refractive index n and thickness ℓ will be attenuated and phase-shifted so that at the output the field will be given by

$$\mathcal{E}_{\text{out}} = \mathcal{E}_{\text{in}}e^{inkz} = \mathcal{E}_{\text{in}}e^{-n_I kz}e^{in_R kz}, \tag{2.23}$$

where $n_{R,I}$ are the real and imaginary parts of the refractive index, $k = 2\pi/\lambda$ is the wavenumber with λ the transition wavelength. The output intensity (the measurable quantity) is then

$$I_{\text{out}} = |\mathcal{E}_{\text{out}}|^2 = I_{\text{in}}e^{-2n_I kz}, \tag{2.24}$$

which is just the Beer-Lambert law with the absorption coefficient $\alpha = 2n_I k$. The frequency dependent transmission is then $\mathcal{T} = I_{\text{out}}/I_{\text{in}}$.

2.3 Motion and the Doppler Effect

So far we have only dealt with stationary atoms. In any room temperature ensemble, the atoms have thermal velocity v, typically around 250 m/s for room temperature atoms. The frequency of light absorbed by the atom is therefore Doppler shifted according to the velocity of the atom, by an amount

$$\Delta_{\text{D}} = -\mathbf{k}\cdot\mathbf{v} = -kv_z, \tag{2.25}$$

where $\mathbf{k}$ is the wavevector of the incoming (monochromatic) laser beam, taken to be along the z-axis. We see immediately from this relation that it is only when there is a component of velocity in the axis of the beam v_z that there is any shift in frequency. Since atomic velocities are distributed in all directions, there are a range of velocities which interact resonantly with the laser beam. In one dimension (along the axis of the beam), a Maxwellian distribution of velocities is assumed such that the atomic density for a given velocity class $N(v)$ is given by

$$N(v) = \frac{N_0}{u\sqrt{\pi}}e^{-v^2/u^2}, \tag{2.26}$$

with the most probable velocity $u = \sqrt{2k_{\text{B}}T/m}$ where m is the mass of the atoms and T their absolute temperature, and we have dropped the z subscripts since we are only dealing with one dimension.

Atoms moving away from a blue-detuned laser beam are Doppler shifted back onto resonance and thus absorb the light, and similarly for a red-detuned laser and atoms moving into the beam. The resonance condition is fulfilled wherever $\Delta + kv = 0$. This leads to an inhomogeneous broadening of the absorption line, with a characteristic Doppler width $\delta\omega_{\text{d}}$, which for hot atoms is significantly larger than the natural linewidth. This places a limit on the spectral resolution, such that features separated by less than the Doppler width are not resolved. For ^{85}Rb atoms at room temperature, the Doppler width $\Gamma_{\text{D}} \approx 2\pi \times 500$ MHz, and so the hyperfine structure lines on the D2 line separated by $\sim 2\pi \times 100$ MHz cannot ordinarily be resolved. To obtain sub-Doppler resolution, usual spectroscopic methods involve a second pump beam (such as saturated absorption/hyperfine pumping spectroscopy [10] or polarisation spectroscopy [11, 12]), although sub-Doppler resolution is possible in a nano-cell without a second beam—this will be expanded upon in Chap. 3.

Adding in Doppler broadening to the formulae in Sect. 2.2 is a relatively simple process. We make the following modifications to Eq. (2.22):

$$\Delta \to \Delta - kv$$
$$N \to N_{\text{vc}}(v)$$

We then must take into account all velocities, so we have, as a function of the laser detuning

$$\chi(\Delta, v) \simeq \frac{iN_{\text{vc}}(v)\mu_{21}^2}{\epsilon_0 \hbar(\Gamma_0/2 - i(\Delta - kv))} \tag{2.27}$$

$$\chi(\Delta) \simeq \int \chi(\Delta, v)dv. \tag{2.28}$$

The lineshape that this convolution produces is known as the Voigt profile, and is discussed in the next section.

2.4 Lineshapes: Lorentzians, Gaussians and the Voigt Profile

The lineshape associated with the light absorption of an ensemble of thermal atoms is a Voigt profile. Mathematically, this is the convolution of a Gaussian, from the Doppler broadening, with a Lorentzian, from the homogeneous natural linewidth, Γ_0. Additional sources of broadening (such as dipole–dipole interactions, which will be discussed later) can also be included by further convolving with the appropriate lineshape. For sources of homogeneous (i.e. all atoms see the same effect) broadening, their lineshape is another Lorentzian. In this case we can take advantage of the fact that a convolution of two Lorentzians with widths w_1 and w_2 is another Lorentzian with width $w_1 + w_2$, which makes computation significantly faster since convolution routines are slow and scale badly with the length of the arrays used.

It is important to understand the relative importance of the various broadening mechanisms. In terms of the vapour temperature there are a few regimes of interest, shown in panels (a–c) of Fig. 2.3. For experiments with cold and ultra-cold atoms (panel (a), calculated here for $T = 1$ mK), we expect the atomic motion to have little effect and therefore the Voigt profile is well approximated as a Lorentzian with natural linewidth Γ_0. As temperature increases there is then a regime where neither Lorentzian nor Gaussian approximations fit well. As temperature increases to room-temperature (b), the Doppler effect is significant and near resonance the lineshape can be well approximated by a Gaussian with width Γ_D. Out in the wings of the resonance however, the Lorentzian component still has a measurable effect [13] owing to the relative scaling of the two functions with detuning. This becomes increasingly

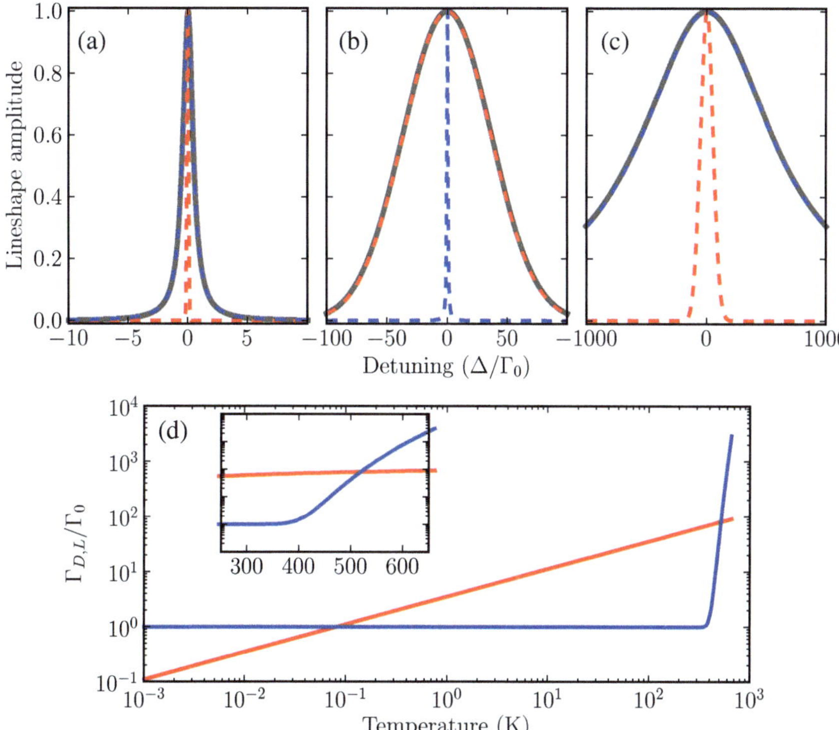

Fig. 2.3 (**a–c**) A comparison between the Voigt profile (*black*) and the Lorentzian (*dashed blue*) and Gaussian (*dashed red*) components. For cold atoms (**a**), the motion of the atoms can be neglected and the Voigt is well approximated by a Lorentzian whose width is the natural atomic linewidth Γ_0. At room temperature (**b**) Doppler broadening dominates and the Voigt is well approximated by a Gaussian, while for temperatures in excess of $\sim$300 °C (**c**) the Lorentzian component due to dipole-dipole interactions dominates (the inset shows an expanded view of the crossing). Panel (**d**) shows the relative widths of the Lorentzian ($\Gamma_L = \Gamma_0 + \Gamma_{dd}$) and Gaussian ($\Gamma_D$) components as a function of temperature, where the near-exponential dependence on temperature of atom-atom interactions can be seen clearly

important as temperature rises further. The contribution to the Lorentzian width from dipole-dipole interactions, Γ_{dd}, sharply increases owing to the linear dependence on number density and hence near-exponential dependence on temperature (see Chap. 5). In this regime ($T \approx 150$–300 °C) we again have a regime where the full Voigt lineshape is approximated badly by both Gaussian and Lorentzian profiles.

The final regime, shown in panel (c), involves very hot ($T \gtrsim 300$ °C) and therefore very dense atomic gases. In this regime, the interactions between atoms dominate over all other line broadening mechanisms and the lineshape can once again be well approximated by only a Lorentzian function with width $\Gamma_L \approx \Gamma_{dd}$. Panel (d) compares the widths of the Lorentzian, $\Gamma_L = \Gamma_0 + \Gamma_{dd}$, and Gaussian Γ_D components as a function of temperature.

2.5 Atomic Structure

It should be clear that for any real system, the above discussion is an idealised model. For the alkali metal atoms, the single valence electron means that the atomic structure is still relatively simple. For the ground state transitions, commonly known as the D1 ($5S_{1/2} \rightarrow 5P_{1/2}$) and D2 ($5S_{1/2} \rightarrow 5P_{3/2}$) lines, there are many allowed transitions between the hyperfine energy levels. The probability amplitude for each transition is found from the dipole matrix element of that transition. In the F basis, with Zeeman sub-levels m_F, the matrix element is $\langle F_g, m_{F_g} | er | F_e, m_{F_e} \rangle$ for a transition between a ground state g and excited state e. In this thesis, we do not deal with applied magnetic fields and neglect the Earth's magnetic field, so we can just consider the transition between F-states. The transition probabilities $|\langle F_g | er | F_e \rangle|^2$ are then the Clebsch–Gordan coefficients, which can be found tabulated for the Rb D1 and D2 lines in [1]. In addition, Rb has two common stable isotopes, ^{85}Rb with a natural abundance of $\approx$72 % and ^{87}Rb with a natural abundance of $\approx$28 %, so this adds a factor of two to the number of transitions.

It is possible to model the full dynamics of this system by solving an $\mathcal{N}$-level version of the master equation (Eq. (2.13)), but this is computationally intensive as there are $\mathcal{N}^2$ coupled equations to solve. The model we use considers only the effect of a weak probe on the system, so optical pumping between levels is negligible. It is therefore assumed that every transition is effectively its own separate two-level system separated in frequency space by the atomic hyperfine structure, and we simply sum over all the dipole-allowed transitions, weighted by their transition probabilities, to calculate the susceptibility at a given laser detuning. This approach has been successful in predicting the full spectrum of the Rb D1 and D2 lines in the past, with agreement between experiment and theory at the 1 % level [1]. Using a more involved model to find the transition amplitudes in the $|m_L, m_S, m_I\rangle$ uncoupled basis has also yielded fruitful results when considering the effect of magnetic fields up to 0.6 T [14–16], with excellent agreement between experiment and theory.

2.6 Summary

The development of a simple model that accurately reproduces experimental spectra has been key to a great deal of research in the Durham atomic and molecular physics group. It is not only a useful tool, in terms of fitting data and extracting parameters, but also gives an ability to predict novel effects. Implementing this principle for very high magnetic fields has led to the development of an atom-based optical isolator [16]; a dichroic beamsplitter based on the Faraday effect [4]; and a tool for off-resonance laser locking at arbitrary frequencies [5].

One of the goals for this work was to extend this model to the subtleties and extra complications introduced by the confinement of atoms in nano-cells, which will be the subject of the next chapter.

References

1. P. Siddons, C.S. Adams, C. Ge, I.G. Hughes, Absolute absorption on rubidium d lines: comparison between theory and experiment. J. Phys. B **41**, 155004 (2008)
2. P. Siddons, N.C. Bell, Y. Cai, C.S. Adams, I.G. Hughes, A gigahertz-bandwidth atomic probe based on the slow-light faraday effect. Nat. Photon **3**, 225 (2009)
3. P. Siddons, C.S. Adams, I.G. Hughes, Optical control of faraday rotation in hot rb vapor. Phys. Rev. A **81**, 043838 (2010)
4. R.P. Abel, U. Krohn, P. Siddons, I.G. Hughes, C.S. Adams, Faraday dichroic beam splitter for raman light using an isotopically pure alkali-metal-vapor cell. Opt. Lett. **34**, 3071 (2009)
5. A.L. Marchant et al., Off-resonance laser frequency stabilization using the faraday effect. Opt. Lett. **36**, 64 (2011)
6. L. Weller, R.J. Bettles, P. Siddons, C.S. Adams, I.G. Hughes, Absolute absorption on the rubidium d1 line including resonant dipoledipole interactions. J. Phys. B **44**, 195006 (2011)
7. S.L. Kemp, I.G. Hughes, S.L. Cornish, An analytical model of off-resonant faraday rotation in hot alkali metal vapours. J. Phys. B **44**, 235004 (2011)
8. C. Tannoudji, J. Dupont-Roc, G. Grynberg, *Atom-Photon Interactions* (Wiley, New York, 1998)
9. J. Gea-Banacloche, Y.-Q. Li, S.-Z. Jin, M. Xiao, Electromagnetically induced transparency in ladder-type inhomogeneously broadened media: theory and experiment. Phys. Rev. A **51**, 576 (1995)
10. D.A. Smith, I.G. Hughes, The role of hyperfine pumping in multilevel systems exhibiting saturated absorption. Am. J. Phys. **72**, 631 (2004)
11. C. Wieman, T. Hänsch, Doppler-free laser polarization spectroscopy. Phys. Rev. Lett. **36**, 1170 (1976)
12. M. Harris et al., Polarization spectroscopy in rubidium and cesium. Phys. Rev. A **73**, 62509 (2006)
13. P. Siddons, C.S. Adams, I.G. Hughes, Off-resonance absorption and dispersion in vapours of hot alkali-metal atoms. J. Phys. B **42**, 175004 (2009)
14. L. Weller, T. Dalton, P. Siddons, C.S. Adams, I.G. Hughes, Measuring the stokes parameters for light transmitted by a high-density rubidium vapour in large magnetic fields. J. Phys. B **45**, 055001 (2012)
15. L. Weller et al., Absolute absorption and dispersion of a rubidium vapour in the hyperfine paschenback regime. J. Phys. B **45**, 215005 (2012)
16. L. Weller et al., Optical isolator using an atomic vapor in the hyperfine paschen-back regime. Opt. Lett. **37**, 3405 (2012)

Chapter 3
Thin Cell Spectroscopy

The purpose of this chapter is to build on the model developed for thermal atoms, introducing the specific features of spectroscopic measurements in the thin cell. There are two main effects due to the thin cell. The first of these is that, because the vapour thickness is shorter than the mean free path of the atoms, there is a partial suppression of the Doppler effect known as Dicke narrowing [1]. Second, the cell geometry is such that etalon effects must be considered, which combined with the variation in the refractive index leads to a large change in the reflective properties across the resonance.

As with the previous chapter, the key is a quantitative model that can be used to make predictions and extract useful parameters.

Experimental details for the data presented in this chapter can be found in Appendix B, which details the optical setup and methods used to calibrate and normalise the data.

3.1 Dicke Narrowing

Dicke narrowing was first proposed in 1953 [1] to explain how collisional processes can result in a narrowing of a Doppler-broadened spectral line. It was first observed in the microwave region soon after [2] but took a further half-century of technological advances to be able to observe the effect in the optical regime, via optical pumping in cells with thicknesses between $10\,\mu$m and 1 mm, using FM spectroscopy [3]. The net result of this effect is that the hyperfine lines that are ordinarily masked by Doppler broadening can start to be resolved, without the use of a counter-propagating pump beam as in conventional sub-Doppler spectroscopy.

When the mean-free-path of the atoms $l = u\tau$, with τ the lifetime of the state and u the most probable thermal velocity, is greater than the length of the cell then the atoms will experience collisions with the walls frequently. From a simple geometry argument, we can estimate how important the wall collisions are. In two dimensions, there is a rectangular excitation region of thickness ℓ and beam waist $2w_0$, which we

J. Keaveney, *Collective Atom–Light Interactions in Dense Atomic Vapours*,
Springer Theses, DOI: 10.1007/978-3-319-07100-8_3,

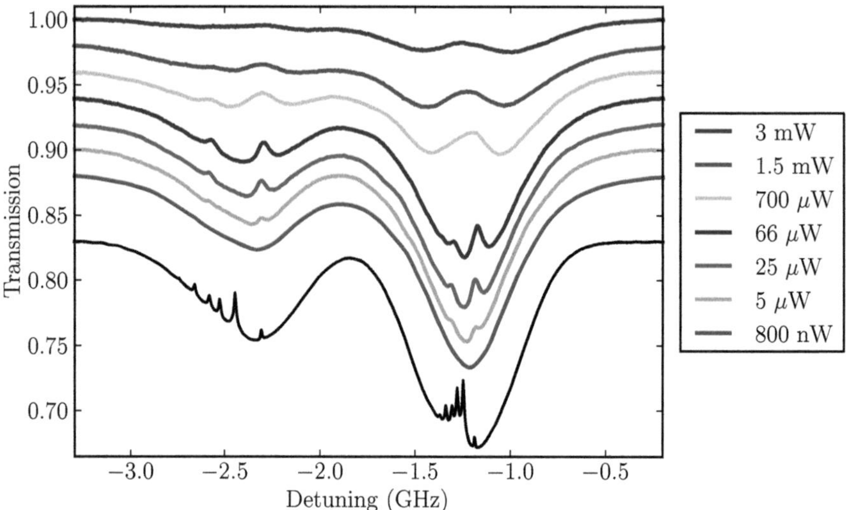

Fig. 3.1 Experimental data for cell thickness $\ell = \lambda$ showing velocity selective optical pumping features for various probe laser powers, as a function of linear detuning $\Delta/2\pi$ over the ^{87}Rb $F_g = 2 \rightarrow F_e = 1, 2, 3$ and ^{85}Rb $F_g = 3 \rightarrow F_e = 2, 3, 4$ D2 resonance lines. At relatively low probe powers the VSOP features are narrow, *centred* around the hyperfine resonances (compare to the pump-probe reference spectrum shown in *black*). As the probe power is increased, there is significant power broadening of the features. An offset of 0.02 has been added between each trace for clarity. Zero detuning represents the weighted line centre of the D2 line

assume forms a hard edge. Assume an atom starts from one of the walls, in the centre of the beam. In order to experience the maximum interaction time, it must make it to the edge of the beam before it hits the other wall. This limits the angle it can make with the wall to $\theta = \arctan(\ell/w_0) \approx \ell/w_0$ (or $\pi - \theta$ in the opposite direction), using the small angle approximation. For a typical geometry, we might have an interaction region with $\ell \sim 200$ nm and $w_0 \sim 10\,\mu$m, which gives $\theta \sim 2 \times 10^{-2}$ rad. Assuming a random uniform distribution of angles, we find only $\theta/(\pi/2) \sim 1\%$ of the atoms experience the maximum interaction time. The atoms that have a small θ will travel more slowly across the cell, and therefore have a smaller Doppler shift.

For a strong laser field, the anisotropic interaction time results in a form of velocity-selective-optical-pumping (VSOP), where only the atoms with a small Doppler shift (the 'slow' atoms) have time to undergo optical pumping. In this case the absorption near resonance is decreased. Figure 3.1 shows experimental data at a cell thickness $\ell = \lambda$ that shows this effect as the probe laser power is increased. For moderate probe powers, the VSOP features are narrow enough that the hyperfine structure is resolved, inside the Doppler background. Hence VSOP can be a useful technique for sub-Doppler spectroscopy with a single laser [4]. For powers below $\sim 1\,\mu$W there is no optical pumping even though the laser is focussed to a waist of $w_0 \approx 10\,\mu$m, corresponding to an intensity greater than the saturation intensity. This will be explored further in Sect. 3.3.

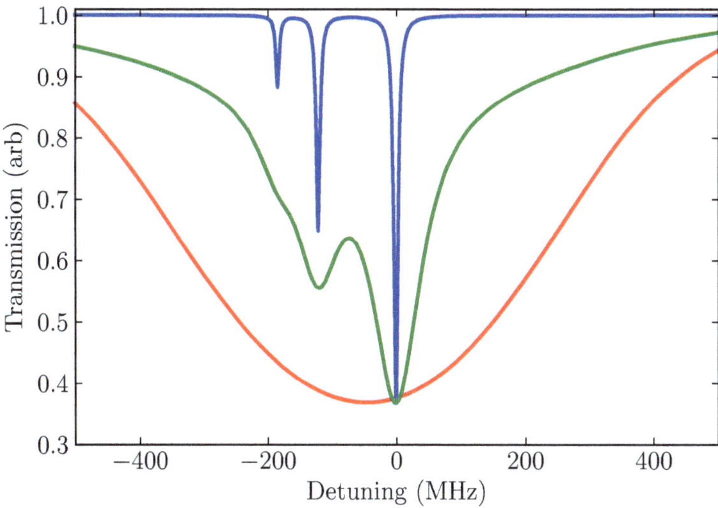

Fig. 3.2 Calculated lineshape comparison between thermal atoms without spatial confinement (*red*), ultracold atoms (*blue*), and thermal atoms confined in a short cell showing the effects of Dicke narrowing (*green*). Calculated as a function of linear detuning $\Delta/2\pi$ in the vicinity of the ^{85}Rb $F_g = 3 \rightarrow F_e = 2, 3, 4$ transitions, with zero detuning set to the $F_g = 3 \rightarrow F_e = 4$ resonance

For a weak probe laser where optical pumping does not occur, the longer interaction time means that the 'slow' atoms experience a higher linear absorption, and results in enhanced sub-Doppler absorption features on top of a Doppler background. Figure 3.2 shows a theoretical comparison of the transmission lineshapes in ultracold atom ensembles (blue), Doppler-broadened thermal ensembles (red) and Dicke-narrowed thermal ensembles (green), calculated for the ^{85}Rb $F_g = 3 \rightarrow F_e = 2, 3, 4$ transitions on the D2 line. The lines are calculations based on realistic experimental parameters and normalised to the same peak absorption. For the case of ultracold atoms, three distinct Lorentzian absorption lines can be observed, each well separated by the hyperfine splitting of the excited state. This is in stark contrast to the Doppler-broadened case, where the lineshape approximates well to a single Gaussian that masks the individual hyperfine features. The Dicke-narrowed case has pronounced sub-Doppler features owing to the partial suppression of the Doppler effect, but the Doppler background is still present in the wings of the resonances.

The narrowing effect also has a coherent component, periodic in the thickness of the cell. It is insightful to consider the superposition principle, where the detected field downstream is the coherent sum of the incident light field and the induced field that is re-radiated by the atoms in the same direction as the laser propagation. This principle becomes especially important when considering interactions between the atoms (see Chap. 5). The key assumption is that collisions between atoms and the wall of the cell reset the system, and cause an immediate decay back to the ground state [5]. The dipolar response is therefore *transient* and starts when the atom is desorbed from the surface and excited (after a relatively long time on the surface [6]). We can thus assume that at one surface, the atom starts in phase with the excitation field. Since only the transient response is important, we do not have the $\pi/2$ phase lag between atom and field, as this is only the steady state oscillator response.

As the cell is so short the time of flight is on the order of a few ns, faster than the natural lifetime of the first excited state (The Rb 5P state lifetime is 27 ns). If there are no other interactions it is therefore likely that the atom will reach the other surface without decaying through spontaneous emission, and the collision with the second surface causes decay with photon emission.

The phase picked up by the laser precesses at the laser frequency ω_p. The atom sees a Doppler-shifted field and thus precesses at its perceived driving frequency $\omega_p + \Delta_D$. A phase difference therefore builds up between atom and field. If the refractive index of the medium is close to unity, valid for low density vapour, [1] then this phase difference is simply given by $e^{i\Delta_D t} = e^{\pm ikvt}$, where v is the component of the atomic velocity in the laser propagation direction. The time t is simply the time it takes to travel from wall-to-wall, $t = \ell/v$, so the phase difference by the time the atom gets to the other wall is $\phi = \pm k\ell$. Significantly, this phase shift is independent of the velocity of the atom—all velocity classes are phase shifted by $\pm k\ell$ when they reach the opposite wall; the sign of the phase shift depends on the direction the atom moves in, i.e. which wall it starts from. Given an uniform distribution of atoms, there is an equal chance of an atom having a phase shift $+k\ell$ as one with $-k\ell$. Their emitted fields will then be $2k\ell$ out of phase with each other. Thus if the cell thickness $\ell = (2n + 1)\lambda/2$, with integer n, then we expect the coherent emission to maximally cancel, whereas for $\ell = n\lambda$ the cancellation does not happen and the spectra return to a Doppler-broadened profile; this periodic 'collapse and revival' was first demonstrated by Dutier et al. [7] in Cs and then by Sarkisyan et al. in ^{87}Rb [4]. For thicknesses greater than $\ell = \lambda/2$ the effect washes out as collisions other than with the walls become more likely. However, these enhanced narrow features have been observed out to a thickness $\ell = 9\lambda/4$ [4]. In the same work, Sarkisyan et al. also showed that the thickness dependence exists only for transmission spectroscopy, and not in fluorescence. One might expect this intuitively, as the transmitted field comes from a coherent process, whereas the off-axis fluorescence is incoherent.

3.1.1 Modelling Dicke Narrowing

The model used to describe Dicke narrowing in this work is phenomenological. We model the number density of atoms with a given velocity v as a bimodal distribution of two Gaussians

$$N_{vc}(v) = CN\left(A\exp(-v^2/u^2) + (1-A)\exp(-v^2/(su)^2)\right) \qquad (3.1)$$

$$= C\left(A\,G_{\text{fast}} + (1-A)\,G_{\text{slow}}\right), \qquad (3.2)$$

---

[1] For high density vapour the index significantly varies across the resonance (see Chap. 6). However, the lifetime of the state reduces significantly due to atom-atom interactions, to the point where a wall-to-wall transit is unlikely. In this case, Dicke narrowing becomes less important to the lineshape.

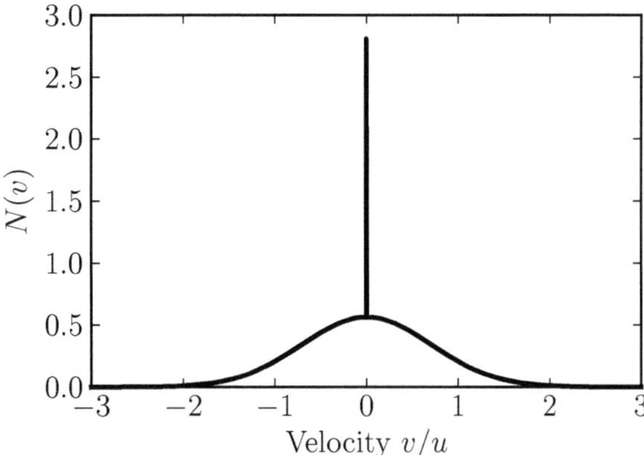

Fig. 3.3 Example velocity distribution based on Eq. (3.2) with $A = 0.2$, $s = 10^{-4}$. The contribution of zero-velocity class atoms is greatly enhanced, simulating the Dicke narrowing features we observe in experimental spectra

where $C = (Au\sqrt{\pi} + (1 - A)su\sqrt{\pi})^{-1}$ is a normalisation constant that ensures $\int N_{vc}(v)dv = N$. The first term is the contribution from 'fast' atoms that make up the usual Maxwellian distribution (G_{fast}), weighted by the coefficient A and with most probable velocity u, and the second term is the narrow Gaussian (G_{slow}) weighted by $(1 - A)$ and with most probable velocity su, where s is a narrowing factor for the 'slow' atoms. An example velocity distribution using this model is shown in Fig. 3.3, with parameters $A = 0.2$, $s = 10^{-4}$. When the Voigt profile is taken into account, there are effectively two convolutions, $G_{\text{fast}} \otimes L$ and $G_{\text{slow}} \otimes L$, where L is the Lorentzian component. So long as the width of the 'slow' distribution is much smaller than the Lorentzian component, we can approximate the convolution as between the Lorentzian and a delta function. Hence the exact value of s is relatively unimportant, so long as it is small enough (we use $s = 10^{-4}$ in the subsequent analysis, but the Gaussian form is retained for ease of normalisation). We use A as a fit parameter to the low-density spectra, after which it is fixed for a given cell geometry. As the density is increased and the Lorentzian component becomes more dominant (see Fig. 2.3), the velocity distribution becomes less important to the overall lineshape.

Figure 3.4 shows example spectra on the D1 and D2 lines which clearly show the effects of Dicke narrowing. Panel (a) shows the D1 spectrum, with a vapour thickness $\ell = 320$ nm and temperature $T = 210\,°C$, while panel (b) shows the narrowing effect on the D2 line, with $\ell = 390$ nm and $T = 190\,°C$. The red lines in the plot are fits to the experimental data (black), which allows the parameter A to vary, along with the temperature and a line broadening parameter. Compared with a Doppler-broadened spectra (grey), the thin cell spectra are much narrower, allowing for resolution of the individual excited state hyperfine levels. These are resolved more for the D1 line due to the larger hyperfine splitting of the $5P_{1/2}$ excited state, and because there are fewer transitions between individual hyperfine levels (8 for D1, as compared with 12 for D2).

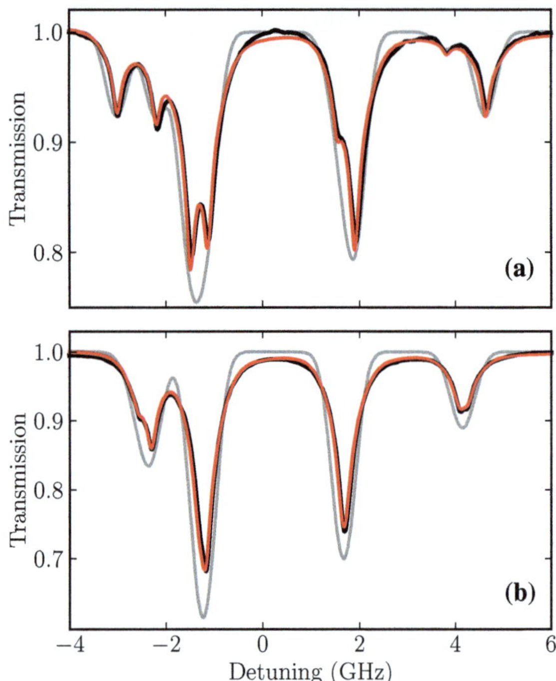

Fig. 3.4 Dicke narrowed spectra as a function of linear detuning $\Delta/2\pi$ over (**a**) the D1 line with a cell thickness $\ell = 320$ nm and temperature $T = 210\,°$C, and (**b**) the D2 line, with $\ell = 390$ nm and $T = 190\,°$C. *Black lines* are experiment, *red* are fits to the data using the model, and the *grey lines* show the Doppler-broadened lineshapes that would be observed in a longer cell. The experimental lines are clearly much narrower than an equivalent Doppler-broadened profile, allowing resolution of the excited state hyperfine levels

3.2 Mixing of Reflection and Transmission

Even though the nano-cell has a wedge profile allowing the local vapour thickness to be tuned (see Appendix A), the angle between the windows is $\sim$0.1 mrad so we can consider them effectively parallel. Because of this, the cell acts as a low-finesse cavity,[2] so etalon effects become important in the transmitted and reflected lineshapes. This effect has been previously observed and a theoretical model developed [9], based on mixing in elements of the lineshape that appears in selective reflection (SR) spectroscopy [10, 11] into the transmitted lineshape, and vice versa. In this section we show that it is possible to predict this effect from classical thin film theory.

Example spectra of simultaneously measured transmission and SR signals around the D2 resonance are shown in Fig. 3.5 for a range of cell temperatures. In this case $\ell = 150$ nm, where the atom-surface interactions broaden the spectral features (see Chap. 4), so even at the lowest temperature the hyperfine excited states are not resolved. The mixing of transmission and SR signals manifests as an asymmetry in the measured lineshapes: in transmission the asymmetry is in the detuning axis (i.e. a difference between the positive-detuned side and the negative-detuned side); in SR

---

[2] Whose finesse, $\mathcal{F}$, is related to the reflectivity, R, of the windows by $\mathcal{F} = \pi\sqrt{R}/(1 - R)$ [8]. For sapphire, $n = 1.8$ and $R = 0.28$, yielding a finesse $\mathcal{F} \sim 2$.

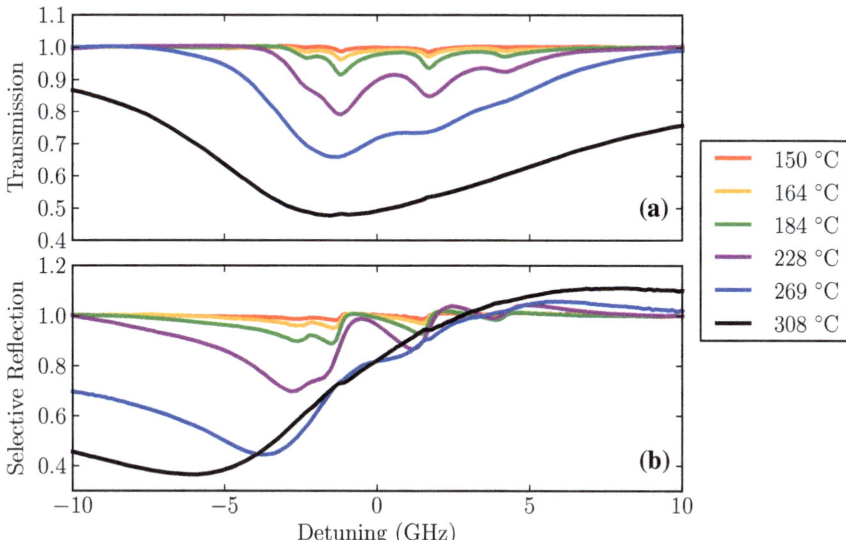

Fig. 3.5 Experimental data showing transmission (**a**) and selective reflection (**b**) signals measured simultaneously on the D2 line for a cell thickness $\ell \approx 150$ nm at various temperatures, showing the mixing of the two signals into each other as a consequence of cavity effects. In the data the effect manifests as an asymmetry for each line, which is most obvious at high temperature. In selective reflection the asymmetry is in the signal axis, whereas in transmission the asymmetry is over the detuning axis. Small VSOP features are present on the data at 308 °C, which are masked by Dicke narrowing at lower temperatures

the asymmetry is in the signal axis, as the spectrum is no longer centred around the far-detuned value (which is set to 1 in our analysis).

For thin layers the SR spectrum does not provide any advantages over transmission spectroscopy, since they are intimately linked by the etalon nature of the cell. When the layer is thicker, however, one may expect the medium to be completely opaque to conventional transmission spectroscopy. In this case SR spectroscopy is a useful technique, since it only probes those atoms that are near to the surface, and has been used extensively to extract shifts and widths of atomic resonances [12–14].

3.2.1 Thin Film Theory

We wish to calculate the reflection of light at the boundary between layers, which are shown schematically in Fig. 3.6. In principle, for an input wave normally incident to an interface made up from many separate layers, the reflection coefficient at the first interface depends on all the layers behind, as there are waves travelling backwards from each interface which must be taken into account. However, as the thickness of the sapphire is much larger than the vapour thickness, then we can regard the Sapphire-Rb interface (ABC) in isolation from the Air-Sapphire interfaces. Just as

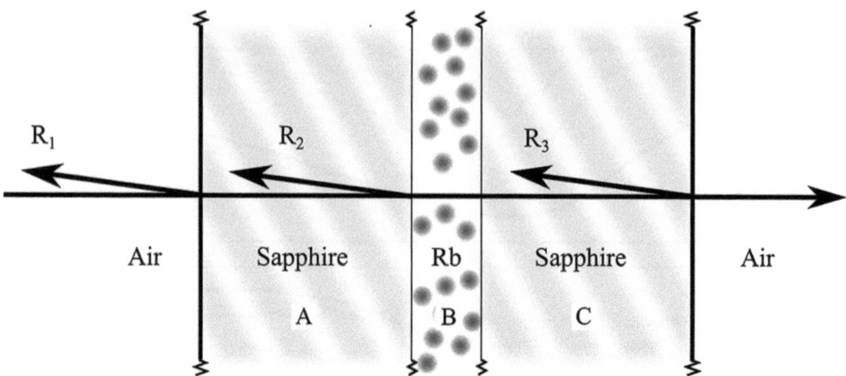

Fig. 3.6 Reflections from each of the boundaries between layers. The thickness of sapphire ($\sim$2 mm) is much larger than the thickness of the Rb vapour layer, which is of the order of the transition wavelength. For this reason, the reflections at the sapphire-Rb interfaces must be considered together, and can be described using thin film theory. The reflection angles have been accentuated for clarity—in the experiment the beam is close to normal incidence

in electrical pathways, the reflectivity at a given boundary depends on the impedance mismatch between the two sides of the boundary.

The impedance of a medium is defined as the ratio between electric and magnetic field components, $Z = E_x/H_y = (\mu_r\mu_0/\epsilon_r\epsilon_0)^{1/2}$, for a wave propagating in the z-direction with the electric field polarised along x [15]. For a non-magnetic medium this reduces to $Z = Z_0/n$ with $Z_0 = (\mu_0/\epsilon_0)^{1/2}$.

As in Fig. 3.6, we consider the input wave as travelling from left to right. We need to calculate the 'input' impedance, Z_{in}, that is present at the boundary AB. The reflected intensity is given by

$$R = \left| \frac{Z_{\text{in}} - Z_A}{Z_{\text{in}} + Z_A} \right|^2, \tag{3.3}$$

which depends on Z_A, the impedance of layer A. Following Brooker [15], we consider each layer in turn, working from right-to-left. This is because for the final layer there is only the transmitted field, with no backward-propagating (reflected) field. We therefore start with the impedance of the final layer, Z_C. This is the 'load' impedance at the boundary BC as experienced by a wave travelling through the thin layer B. The input impedance at AB is then calculated from [15]

$$Z_{\text{in}} = \frac{Z_C - iZ_B \tan(k\ell)}{1 - i(Z_C/Z_B)\tan(k\ell)}. \tag{3.4}$$

For the specific case we are concerned with, $Z_A = Z_C = Z_0/n_{\text{sapphire}}$ with $n_{\text{sapphire}} = 1.8$ and $Z_B = Z_0/n(\omega)$ where $n(\omega)$ is the frequency dependent refractive index of the Rb vapour.

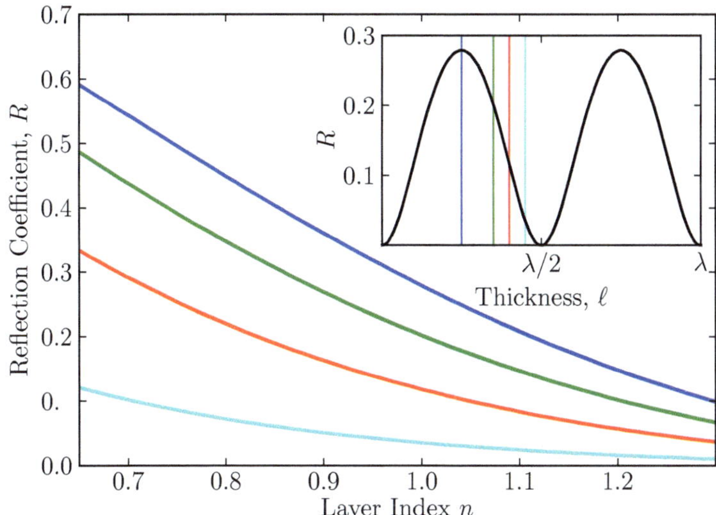

Fig. 3.7 Variation of reflectivity with layer index n for various vapour thicknesses, which are shown as vertical lines in the inset. For thicknesses which are odd multiples of $\lambda/4$ the effect is maximal while for even multiples of $\lambda/4$ the effect disappears completely. (*inset*) For a given layer index (calculated here for $n = 1$), the reflectivity goes through oscillations as the length is varied

Figure 3.7 shows the reflection coefficient as a function of vapour index n and thickness ℓ for a vapour confined between two sapphire plates. For the case of $\ell = \lambda/4$, the reflection coefficient changes by a factor of ~ 3 between $n = 0.8$ and $n = 1.2$. Clearly, this can have a significant effect on the transmitted lineshape. The maximum effect occurs at $\ell = \lambda/4$ (and odd multiples of $\lambda/4$), which can be understood in terms of interference of the light reflected from the two interfaces AB and BC. The phase difference between the two waves reflected here is $2k\ell + \pi$, where the extra π phase shift is due to the reflection at BC as the wave travels into a more dense medium.

It is insightful to see how this alters a single resonance line, before looking at the full spectrum. To this end, we use the results of the previous chapter for the weak-probe susceptibility of a two-level atom, namely

$$\chi(\Delta) = \frac{iK}{\Gamma_0/2 - i\Delta} \,, \tag{3.5}$$

$$n(\Delta) = \sqrt{1 + \chi(\Delta)} \,, \tag{3.6}$$

where K is an amplitude factor. Figure 3.8 shows the refractive index profile, $n(\Delta)$, and the transmitted lineshape in the absence of etalon effects, $\mathcal{T}(\Delta)$ (black lines in panels (a) and (b), respectively). The refractive index profile gives rise to a frequency dependent reflectivity, $R(n(\Delta))$ (red line in panel (a)), which significantly alters the transmitted intensity (red line in panel (b)). The transmitted intensity has been

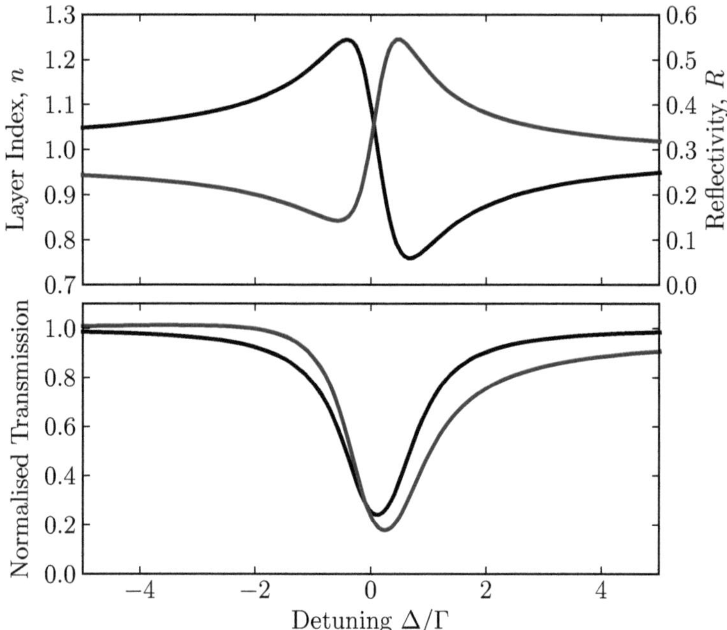

Fig. 3.8 Effect of reflectivity on a single absorption line. **a** The varying refractive index (*black*) over a resonance line gives rise to a large change in the reflectivity (*red*) of a thin layer, calculated for $\ell = \lambda/4$, which in turn affects the transmitted lineshape shown in panel (**b**). The *solid red* (*black*) *line* shows the transmitted intensity with (without) reflectivity effects included, which have been normalised so that when there is no interaction, $n = 1 \implies \mathcal{T} = 1$

normalised so that in the absence of the atomic vapour ($n = 1$), there is unity transmission, consistent with the presentation of experimental data. Considering reflections, the transmitted lineshape $\mathcal{T}_R(\Delta)$ is thus given by

$$\mathcal{T}_R(\Delta) = C\,\mathcal{T}(\Delta) \times (1 - R(n(\Delta)))\,, \tag{3.7}$$

where $C = 1/(1 - R(n = 1))$ is the normalisation constant. This normalisation leads to areas of the spectrum (in the red-detuned wing) where $\mathcal{T}_R > 1$, due to less light being reflected when there are atoms present as compared with an empty cavity, as the refractive indices are more closely matched. The now asymmetric lines also cause an apparent shift of the peak absorption, though this is not a shift of the resonance.

Due to the thickness dependence of the reflectivity, we can directly compare the lineshapes where the etalon effect disappears, for $\ell = n\lambda/2$, and where it is maximal, for $\ell = (2n-1)\lambda/4$. Figure 3.9 shows two sets of data, with equivalent $N\ell$-products. Panel (a) shows data for $\ell = \lambda/4$ while for panel (b) $\ell = \lambda/2$, as the laser is tuned over the D2 resonance. The optical depths are different due to the greater effect of dipole–dipole interactions at higher density, and the apparent enhancement of the dipole–dipole interaction for thicknesses $\ell \leq \lambda/4$ (see Chap. 5). There are twelve

Fig. 3.9 Comparison of reflectivity effects at (**a**) $\ell = \lambda/4$ ($T = 250\,°\mathrm{C}$, $N \approx 5 \times 10^{15}\ \mathrm{cm}^{-3}$) and (**b**) $\ell = \lambda/2$ ($T = 226\,°\mathrm{C}$, $N \approx 2.5 \times 10^{15}\ \mathrm{cm}^{-3}$), as a function of linear detuning $\Delta/2\pi$. The $N\ell$-product is approximately the same in both cases. *Black lines* are experimental data. For $\ell = \lambda/4$ the effect is maximal and there is large asymmetry in the lineshape, which is not seen at $\lambda/2$. The *dashed lines* (which overlaps with the *solid line* in panel **b**) show the predicted lineshape in the absence of reflectivity effects

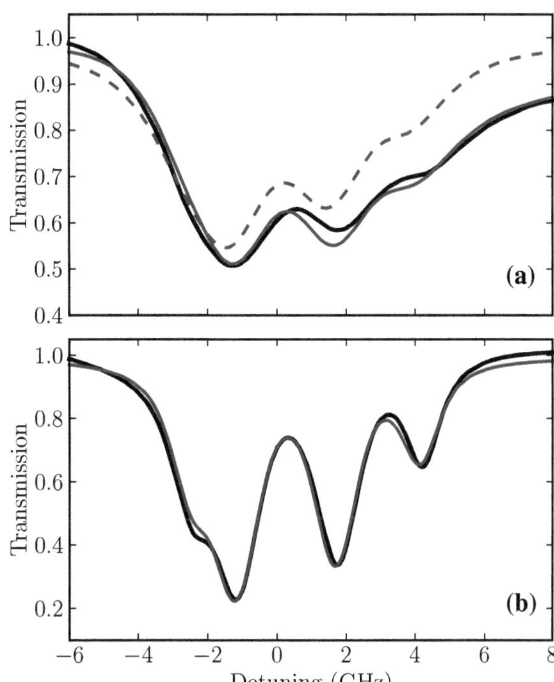

individual hyperfine transitions; three from each of the two hyperfine ground states of the two isotopes. On the figure, the transitions can be identified in groups of three, as the excited state hyperfine splitting (of the order of 100 MHz) is much smaller than the ground state splitting (3 GHz for ^{85}Rb, 7 GHz for ^{87}Rb); the outer two peaks at $\Delta = -2.5$ and 4.5 GHz therefore correspond to the ^{87}Rb lines, whilst the peaks at -1 and 2 GHz correspond to ^{85}Rb. The dashed lines show the calculated lineshape without reflectivity effects, present on both panels but overlapped exactly with the solid line for $\ell = \lambda/2$. The solid lines show the calculated lineshape including reflectivity. At $\ell = \lambda/4$, and especially on the blue-detuned side of the resonance, the data clearly fit better to the theory curve including reflectivity effects, and there is an overall asymmetry to the lineshape that is not present at $\ell = \lambda/2$.

3.3 The Weak-Probe Limit

In many spectroscopic measurements, the aim is to investigate the properties of the medium in question without significantly altering them. In the experiments detailed in the majority of this work, this is achieved by making the laser intensity weak enough that the fraction of atoms excited is very small and the probability of decay via stimulated emission is negligible compared to spontaneous emission.

Conventionally, the saturation intensity is given by [16]

$$I_{\mathrm{Sat}} = \frac{2\pi^2 \hbar c \Gamma_0}{3\lambda^3} . \tag{3.8}$$

The intensities required to be in the 'weak probe limit' depend crucially on the geometry of the system [17, 18] and therefore are sensitive to the atomic interaction time with the laser field. For conventional 'long' (cm and above) cells with unfocussed beams, intensities of the order of $10^{-3} I_{\mathrm{Sat}}$ are required to observe a fully 'weak probe' response [17, 19].

We have already demonstrated in Fig. 3.1 how the anisotropic distribution of interaction times in the thin cell leads to optical pumping for the 'slow' velocity classes which interact the longest with the beam, so we might expect that from the confined geometry in the thin cell the intensity can be larger before optical pumping occurs.

In order to minimise the vapour thickness variation across the beam (see Appendix A), relatively tight focussing is required. This increases the intensity and thus constrains the optical power that can be used, limiting the signal on the detector. From a purely experimental point of view, it is useful to know how much optical power we can use to maximise the signal-to-noise ratio while still avoiding optical pumping.

To compare between cells with different thickness, we require the intensity to be the same. Roughly speaking, we can consider the intensity in the cell almost constant (within a factor of 2) when the cell is shorter than twice the Rayleigh range, which is given by

$$z_R = \frac{\pi w_0^2}{\lambda} , \tag{3.9}$$

and is the distance away from the focus in the propagation axis where the spot size doubles.

For the data presented in Fig. 3.10 we have a focus with a $1/e^2$ radius $w_0 = 15\,\mu$m, which gives $z_R = 3.6$ mm. We therefore compare the peak absorption for a range of cell thicknesses, up to 4 mm (the longest available below the 7.2 mm cutoff). We plot the peak absorption as a function of the laser intensity, normalised to the absorption value for the lowest intensity. For thin vapour cells, the data suggests that the intensity can be orders of magnitude larger than for longer cells while still avoiding optical pumping, since the interaction time is severely restricted by the geometry of the cell. In practice, this means we can use $\sim 1\,\mu$W of optical power focussed to a waist of $\sim 10\,\mu$m in cells with $\ell < 2\,\mu$m before the signal is altered by optical pumping. Another interesting feature of Fig. 3.10 is shown when we look at the difference between open and closed transitions. In this case, the open transition is the absorption measured at the resonance frequency of the ^{85}Rb $F_g = 2 \rightarrow F_e = 3$ transition, and the 'closed' transition is the absorption measured with the laser on resonance with the ^{85}Rb $F_g = 3 \rightarrow F_e = 4$ transition. Since the medium is Doppler broadened, there will be some contributionfrom the other excited state hyperfine levels, so the

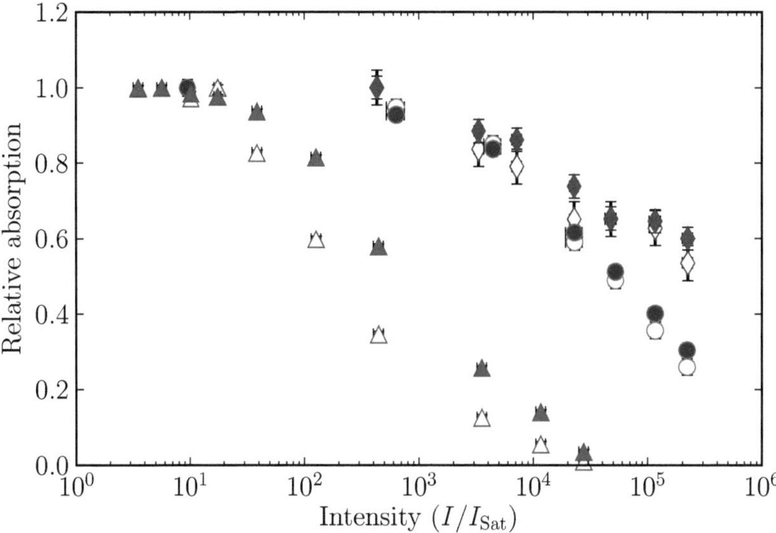

Fig. 3.10 Peak absorption as a function of laser intensity for cell thicknesses of 120 nm (*red diamonds*), 390 nm (*green circles*), and 4 mm (*blue triangles*), probed with a focussed laser beam with a $1/e^2$ radius of 15 μm. The smallest intensities correspond to a power of 50 nW and are already larger than I_{Sat}. The *open* (*closed*) symbols represent the absorption on the *open* (*closed*) transition in ^{85}Rb, $F_g = 2\,(3) \rightarrow F_e = F_g + 1$

'closed' transition is partially open. Despite this, we see a marked difference in behaviour between the two, and in the longer cells the relative difference between open and closed transitions is much larger than for the thin cells. This is probably due to collisions with the cell walls that repopulate the ground states equally, so pumping into a dark ground state is less important in the thin cells than in the longer cells, since the atoms remain in the dark state for much less time.

3.4 Summary

In this chapter, we have explored how in thin vapour cells, a number of effects complicate the lineshapes of the transmitted and reflected spectra. We have extended the model developed in the previous chapter to these thin vapours, and in the next chapters we will apply this model to extract data from experimental spectra about the interatomic and atom-surface interactions.

References

1. R. Dicke, The effect of collisions upon the Doppler width of spectral lines. Phys. Rev. **89**, 472 (1953)
2. R. Romer, R. Dicke, New technique for high-resolution microwave spectroscopy. Phys. Rev. **99**, 532 (1955)
3. S. Briaudeau, S. Saltiel, G. Nienhuis, D. Bloch, M. Ducloy, Coherent Doppler narrowing in a thin vapor cell: Observation of the Dicke regime in the optical domain. Phys. Rev. A **57**, R3169 (1998)
4. D. Sarkisyan et al., Spectroscopy in an extremely thin vapor cell: Comparing the cell-length dependence in fluorescence and in absorption techniques. Phys. Rev. A **69**, 065802 (2004)
5. S. Briaudeau, D. Bloch, M. Ducloy, Detection of slow atoms in laser spectroscopy of a thin vapor lm. Europhys. Lett. **35**, 337 (1996)
6. H. de Freitas, M. Oria, M. Chevrollier, Spectroscopy of cesium atoms adsorbing and desorbing at a dielectric surface. Appl. Phys. B **75**, 703 (2002)
7. G. Dutier et al., Collapse and revival of a Dicke-type coherent narrowing in a sub-micron thick vapor cell transmission spectroscopy. Europhys. Lett. **63**, 35 (2003)
8. W.T. Silfvast, *Laser Fundamentals*, 2nd edn. (Cambridge University Press, Cambridge, 2004)
9. G. Dutier, S. Saltiel, D. Bloch, M. Ducloy, Revisiting optical spectroscopy in a thin vapor cell: Mixing of renection and transmission as a Fabry Perot microcavity effect. J. Opt. Soc. Am. B **20**, 793 (2003)
10. J. Woerdman, M.F.H. Schuurmans, Spectral narrowing of selective reflection from sodium vapour. Opt. Commun. **14**, 248 (1975)
11. M.F.H. Schuurmans, Spectral narrowing of selective reflection. J. Phys. (Paris) **37**, 469 (1976)
12. J.J. Maki, M.S. Malcuit, J.E. Sipe, R.W. Boyd, Linear and nonlinear optical measurements of the Lorentz local field. Phys. Rev. Lett. **67**, 972 (1991)
13. V.A. Sautenkov, H. van Kampen, E. Eliel, J. Woerdman, Dipole-Dipole broadened line shape in a partially excited dense atomic gas. Phys. Rev. Lett. **77**, 3327 (1996)
14. H. Li, V.A. Sautenkov, Y.V. Rostovtsev, M.O. Scully, Excitation dependence of resonance line self-broadening at different atomic densities. J. Phys. B **42**, 065203 (2009)
15. G. Brooker, *Modern Classical Optics* (Oxford University Press, Oxford, 2003)
16. R.W. Boyd, *Nonlinear Optics*, 3rd edn. (Academic Press, New York, 2008)
17. B.E. Sherlock, I.G. Hughes, How weak is a weak probe in laser spectroscopy? Am. J. Phys. **77**, 111 (2009)
18. T. Lindvall, I. Tittonen, Interaction-time-averaged optical pumping in alkali-metal-atom Doppler spectroscopy. Phys. Rev. A **80**, 032505 (2009)
19. P. Siddons, C.S. Adams, C. Ge, I.G. Hughes, Absolute absorption on rubidium D lines: comparison between theory and experiment. J. Phys. B **41**, 155004 (2008)

Chapter 4
Atom–Surface Interactions

A direct consequence of using cells that are shorter than the atomic transition wavelength is that atom-surface interactions can become important. In this chapter we investigate these interactions in the thin cell. We show that even for the first excited state, there is a significant interaction when the cell thickness is of the order of the reduced wavelength, $\lambda/2\pi$. This interaction leads to both a broadening and a shift of the resonance lines.

4.1 The van der Waals Atom–Surface Interaction

The van der Waals (vdW) interaction between two bodies is one of the most important interactions in nature, responsible for the chemical bond of the same name, a correction to the ideal gas law (the van der Waals equation [1]) and the non-resonant interaction between a pair of identical atoms, discussed in the next chapter and recently measured directly using two trapped Rydberg atoms [2].

The van der Waals interaction also exists between an atom and a nearby dielectric surface, and acts to shift and broaden the atomic spectral features. We consider the atom as an instantaneous dipole, as shown schematically in Fig. 4.1. Near a surface, the dipole induces a local charge distribution within the partially reflecting surface mirroring the atom's own dipole moment [3]. This induced dipole in the surface interacts with the atomic dipole in the same way as between two atomic dipoles [4]. In the short-range regime, this interaction produces a potential $\Delta_{\mathrm{vdW}} = -C_3/r^3$, where the C_3 coefficient depends on the surface, and the atomic state. A positive C_3 coefficient implies an attractive interaction. Further away from the surface, retardation effects cause the scaling to drop to $1/r^4$, known as the Casimir-Polder interaction [5], whilst at very large separations, thermal fluctuations in the electromagnetic field change the interaction back to a $1/r^3$ potential [6]. The larger the polarisability of the atomic state, the stronger the interaction with the surface,

J. Keaveney, *Collective Atom–Light Interactions in Dense Atomic Vapours*, 35
Springer Theses, DOI: 10.1007/978-3-319-07100-8_4,
© Springer International Publishing Switzerland 2014

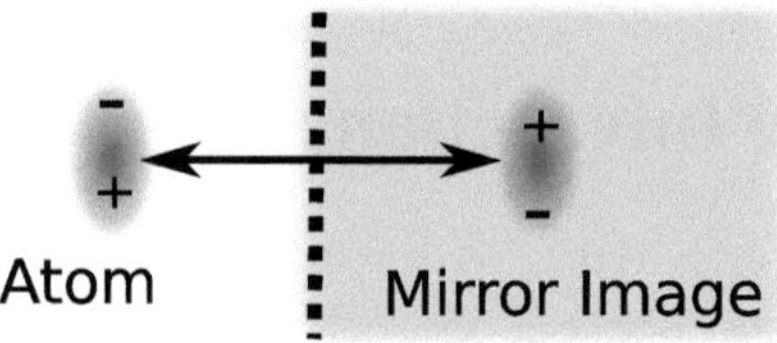

Fig. 4.1 Schematic of the dipolar interaction with a surface. A dipole near a surface locally polarises it, so the real dipole sees a mirror image of itself in the surface. The interaction is attractive, leading to a red-shift of the atomic energy levels

leading many experiments to use Rydberg atoms to investigate these interactions [2, 3, 7].

In most cases the atom-surface interaction is attractive, leading to a red-shift (shift to lower frequency) of the atomic levels [8]. A notable exception to this is when a resonance in the surface (known as a surface polariton [9]) coincides with an atomic transition—in this case, the resonance can lead to a blue-shift (shift to higher frequency) of the atomic lines [10]. There have been extensive studies into the surface polariton resonances of sapphire glass (which is used in this experiment) with Cs [11] and Rb vapours [10]. Surface polaritons are not an issue in our experiment, as the resonances between Rb and sapphire are at lower wavelengths which are only accessible via transitions from higher excited states than the $5P_{3/2}$ level [10].

The shift of the spectroscopic line is in principle the difference in the shifts of the two states, the ground state $5S_{1/2}$ and the excited state $5P_{3/2}$ [12]. Previous experiments using sapphire in conjunction with the low-lying excited states of alkali metals have reported a spectroscopic C_3 coefficient of 2 kHz μm^3 for the $6S_{3/2}$–$6P_{3/2}$ transition in Cs [12].

4.1.1 Atom–Surface Potential in the Nano-Cell

Between the two sapphire walls of the nano-cell, a Rb atom sees two potentials—one from each wall. For a cell of thickness ℓ, the energy shift is of the form

$$\Delta_{\mathrm{vdW}} = -\frac{C_3}{z^3} - \frac{C_3}{(\ell - z)^3}, \tag{4.1}$$

and is plotted in Fig. 4.2 in scaled units. At the centre of the cell, $z = \ell/2$, the total shift is $\Delta_{\mathrm{vdW}} = -16C_3/\ell^3$. We include this potential into the susceptibility model presented in the previous chapter by assuming an average response, with a single line shift parameter Δ_{vdW} and contribution to the linewidth Γ_{vdW} that is incorporated into the fitting (see Chap. 5). For ease of implementation we account for the increase in linewidth in the same way as for self-broadening; an additional contribution to the Lorentzian component. It should be noted that this is without any

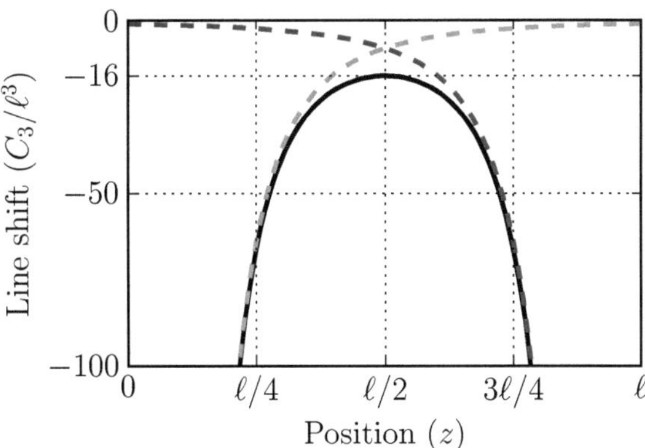

Fig. 4.2 Atom-surface potential in the nano-cell, calculated from Eq. (4.1). Both walls contribute a $-C_3/r^3$ potential

physical significance—the actual modification to the lineshape is an inhomogeneous broadening caused by the spatial anisotropy of the surface interaction. Hence to model properly the atom-surface interaction, one needs to consider the transit of the atoms across the cell and integrate the anisotropic surface interaction they experience as a function of distance across the cell, as in Ref. [13]. Unlike Doppler broadening, where the symmetric velocity distribution leads to a symmetric broadening of the resonance line, the shifts from the atom-surface interaction are only to negative frequencies, leading to an asymmetric lineshape with a longer negative frequency tail, as noted in Ref. [12].

4.2 Monte-Carlo Simulation of Interaction Time Distribution

Whilst fitting the experimental data using a single shift parameter provides an upper bound estimate of the C_3 coefficient, this assumes that all the atoms interact with the walls only in the centre of the cell. Clearly this is a naive assumption, and therefore can be improved by inferring the distribution of atoms in the cell.

We estimate the distribution of atomic interaction times by employing a Monte-Carlo (MC) simulation, where the atoms are assumed to start in random positions on the edges of the interaction region. We assume this interaction region is a box with hard edges whose dimensions are determined by the beam waist and cell thickness, and move at a speed $u = \sqrt{2k_B T/m}$ in a random direction, uniformly distributed over the 4π solid angle. The atoms are assumed to move ballistically—no interactions are included.

Since we investigate the interaction using CW spectroscopy, it is known (Chap. 3) that there is an anisotropic distribution of interaction times due to the cell geometry. The spectroscopy is most sensitive to those atoms that interact for longest, so we only count those atoms that traverse the whole of the interaction region, i.e. those atoms that do not hit the cell windows. These are the same 'slow' (small Doppler shift)

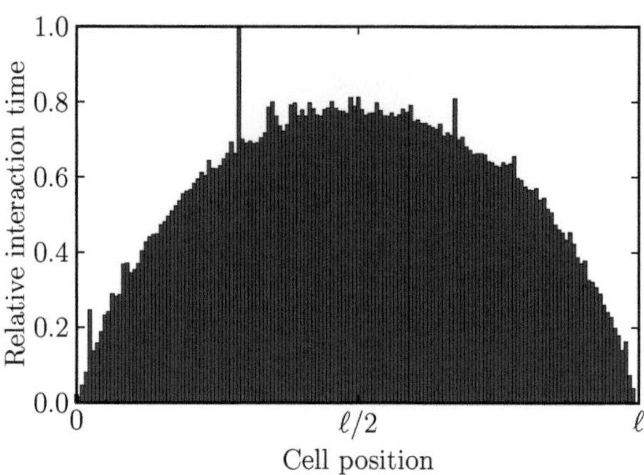

Fig. 4.3 Histogram of atomic interaction times as a function of position in the cell, calculated from the Monte-Carlo model detailed in the main text. From this we can infer a distribution of shifts

atoms that are modelled with the bimodal velocity distribution, so this MC method gives us an estimate of how many 'slow' atoms there are compared to the amount that we attribute to the full Doppler average.

The cell is split into segments with a width in the propagation direction δz. The time the i-th atom spends traversing the region between position z and $z+\delta z$ depends on the component of its velocity in the z-direction, v_z. Away from the edges this time is just $\delta z/v_z$. The total time that the atoms spend in any region is then summed over $\mathcal{N} = 10^8$ random positions and trajectories. In Fig. 4.3 we plot the result of the simulation, in this case calculated for a cell thickness $\ell = 90$ nm. The distribution is approximately parabolic, symmetric about the centre of the cell. From the simulation we find the fraction of atoms that interact fully, without hitting the walls, is approximately 0.02 %, which roughly agrees with the simple geometric argument in Sect. 3.1.

The relative interaction times are used as a weighting factor, such that the observed shift is the sum of the shifts at the centre of each histogram bin weighted by the relative interaction time per bin, normalised such that the sum of all interaction times equals 1. Compared to the approximation that all the atoms see the shift at the centre of the cell, where the shift would be $-16C_3/\ell^3$, the weighted sum differs by a factor of 0.48. The variation in this value across different runs (with different random seeds) is less than 0.001.

4.3 Experiment

Transmission spectroscopy is utilised to extract line shifts and widths; most of the experimental details are documented in Appendix B. An optical power of 300 nW focussed to a $1/e^2$ radius of approximately 30 μm was used for these measurements, leading to an variation in the thickness of the cell over the laser spot of around 5 nm. All data presented here were taken at a constant temperature of $T = 195$ °C which corresponds to an atomic number density $N = 7 \times 10^{14}$ cm^{-3}. In terms of linewidth

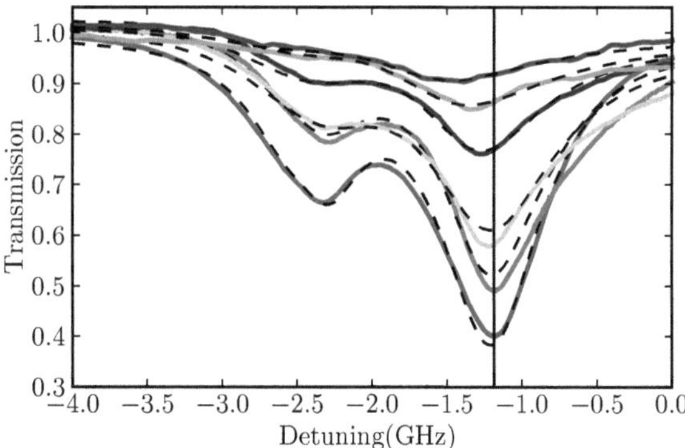

Fig. 4.4 Experimental D2 transmission spectra and fit to the susceptibility model (*black dashed lines*) at a temperature $T = 195\,°C$, corresponding to a number density $N = 7 \times 10^{14}\,\text{cm}^{-3}$, for various cell thicknesses: $\ell = 390\,\text{nm}$ (*purple*, x1); $\ell = 195\,\text{nm}$ (*blue*, x2.5); $\ell = 110\,\text{nm}$ (*cyan*, x5); $\ell = 90\,\text{nm}$ (*green*, x7.5); $\ell = 70\,\text{nm}$ (*orange*, x10); $\ell = 50\,\text{nm}$ (*red*, x15). The data have been scaled by the factor in brackets for clarity. Zero detuning is the weighted line centre of the D2 line

contribution from dipole-dipole interactions this corresponds to an additional collisional broadening $\beta N \approx 2\pi \times 70\,\text{MHz}$; the total expected shift is smaller than this, the exact value depends on the thickness of the medium (see Chap. 5).

In Fig. 4.4 we show the experimental data (coloured lines) with the fitted model (black dashed lines). The data are fit only in the region shown on the plot. The signal decreases as thickness is decreased due to decreased optical depth and hence signal-to-noise worsens (there is less than 1 % absorption for the smallest two thicknesses), leading to larger errors in the fit parameters. This also means that using transmission spectroscopy at this density, data for the thinnest regions of the cell are inaccessible due to lack of absorption. The thinnest region we can practically investigate is 50 nm. Despite the optical depth limitations, we can still extract parameters from the fits. The line shifts extracted from fitting the data are shown in Fig. 4.5. We fit these points to a function of the form K/ℓ^3 and extract the coefficient $K = -62 \pm 9\,\text{kHz}\,\mu\text{m}^3$. By assuming all of the atoms are in the centre of the cell, where they experience the smallest shift, we can place an upper bound on the C_3 coefficient for a single wall interaction. In this case we have $C_3 \leq K/16 = 3.9 \pm 0.5\,\text{kHz}\,\mu\text{m}^3$. A better estimate can be gained from the spatial distribution of atomic interaction times, which was calculated with the Monte-Carlo method in the previous section. Using this weighting, we estimate a spectroscopic shift coefficient $C_3 = 1.9 \pm 0.3\,\text{kHz}\,\mu\text{m}^3$, similar to previous measurements on the Cs $6S_{1/2}$ to $6P_{3/2}$ transition [12].

In Fig. 4.6 we plot the extracted line widths. Two sets of data are plotted: the open symbols are the widths extracted directly from the fit parameters from the data in Fig. 4.5; the closed symbols represent the zero-density offsets from fitting the density-dependent data presented in Chap. 5. The two pseudo-independent methods give similar results. Fitting the two methods separately to a an inverse power law with a floating exponent p, we obtain $p = 1.0 \pm 0.2$ for the widths extracted from

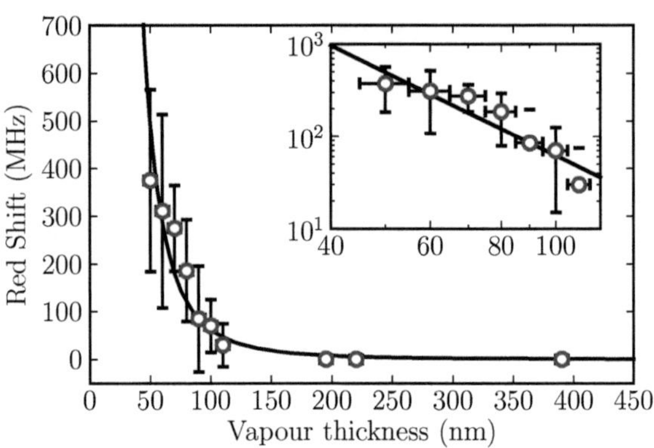

Fig. 4.5 Extracted shifts due to vdW atom-surface interaction. The line is a fit to the data with a ℓ^{-3} form. The inset shows the same data on a log scale

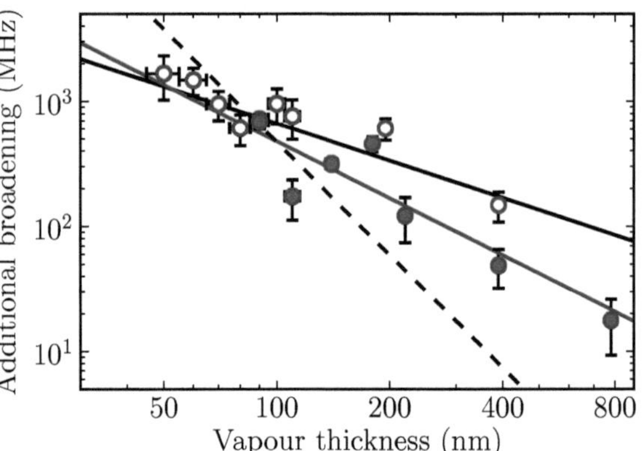

Fig. 4.6 Atom-surface broadening parameter extracted from fitting the data in Fig. 4.4 (*open symbols*), and from the zero-density offset of the data presented in Fig. 5.9. The *solid lines* are fits to an inverse power law, $\Gamma_{\text{off}} = K \times \ell^{-p}$, with exponents $p = 1.5 \pm 0.4$ (*red*) and $p = 1.0 \pm 0.2$ (*black*). The *dashed black line* is a fit to an inverse-cubed function

the individual fits, whilst an exponent $p = 1.5 \pm 0.4$ is obtained for the zero-density offset data. Neither data set fit well to an inverse-cubed function (black dashed line). That these broadening parameters do not fit to an inverse-cubed curve is not unexpected, confirming that the Lorentzian broadening model is a bad approximation. It is therefore somewhat surprising that the model still matches the experimental spectra as well as it does.

For small thicknesses, accurate determination of the line shift and width is difficult due to the low signal-to-noise, limited by the overall amount of absorption. An improvement to the current experiment which is relatively easy to implement is to detect the off-axis fluorescence as the laser is tuned across the resonance. By using a single-photon counting module, we can detect very small fluorescence intensity, so the probe beam can remain weak. In principle, aside from an overall background level (which should be uniform over the laser scan and can be subtracted in postprocessing), the signal lies on a zero background, and therefore the signal-to-noise for situations in which there is very little absorption can be vastly improved. Preliminary work has been carried out on this—further details can be found in Appendix C of this thesis; Fig. C.4 shows a fluorescence spectrum.

4.4 Outlook

We have shown how the atom-surface interaction shifts and broadens the spectral lines as the cell thickness is reduced. The shifts which we extract from a model of the weak-probe susceptibility follow the expected $1/\ell^3$ behaviour, allowing an estimate of the C_3 coefficient which characterises the strength of the interaction between the sapphire surface and a Rb atom in the $5P_{3/2}$ state. Using the measured shifts in combination with a Monte-Carlo simulation to estimate the interaction time spatial distribution, we estimate the atom-surface interaction strength to be $C_3 = 1.9 \pm 0.3$ kHz $\mu\mathrm{m}^3$. The current model uses a bad approximation for the atom-surface contribution to the spectral line widths. Including a full treatment of the corresponding inhomogeneous broadening caused by the surface interaction would improve the theoretical model, and will be a subject of future work.

In extremely thin vapours at relatively low density where the amount of absolute absorption is small, fluorescence measurements yield better signals than transmission spectroscopy (see Appendix C) and hence should allow for more precise determination of line shifts. Specifically, it would allow access to the thinnest regions of the cell where currently transmission spectroscopy is limited by poor signal.

As density is increased, the atom-surface interaction becomes less important to the overall system, as the dipole-dipole interactions between identical atoms eventually dominate. As a comparison, for a cell thickness of 90 nm, we observe an atom-surface shift of around 100 MHz, as compared to 2 GHz at the highest density due to dipole-dipole interactions. Similarly for the broadening, the dipole-dipole interaction eventually becomes around an order of magnitude higher than the atom-surface broadening at the highest atomic densities, as we will see in the next chapter.

References

1. H.D. Young, R.A. Freedman, *University Physics*, 12th edn. (Pearson Education, New Jersey, 2008)
2. A. Vernier, R. Chicireanu, T. Lahaye, A. Browaeys, Direct measurement of the van der Waals interaction between two single atoms. arXiv.org, 1302.4262 (2013). arXiv:1302.4262v1.
3. V. Sandoghdar, C.I. Sukenik, E.A. Hinds, S. Haroche, Direct measurement of the van der Waals interaction between an atom and its images in a micron-sized cavity. Phys. Rev. Lett. **68**, 3432 (1992)
4. I. Hamdi et al., Laser spectroscopy with nanometric gas cells: distance dependence of atom surface interaction and collisions under confinement. Laser Phys. **15**, 987 (2005)
5. C.I. Sukenik, M.G. Boshier, D. Cho, V. Sandoghdar, E.A. Hinds, Measurement of the Casimir-Polder force. Phys. Rev. Lett. **70**, 560 (1993)
6. M. Antezza, L.P. Pitaevskii, S. Stringari, Effect of the Casimir-Polder force on the collective oscillations of a trapped Bose–Einstein condensate. Phys. Rev. A **70**, 053619 (2004)
7. V. Sandoghdar, C. Sukenik, S. Haroche, E. Hinds, Spectroscopy of atoms confined to the single node of a standing wave in a parallel-plate cavity. Phys. Rev. A **53**, 1919 (1996)
8. A. Aspect, J. Dalibard, Measurement of the atom-wall interaction: from London to Casimir-Polder. Séminaire Poincaré **1**, 67 (2002)

9. J. Nkoma, R. Loudon, D. Tilley, Elementary properties of surface polaritons. J. Phys. C **7**, 3547 (1974)
10. H. Failache, S. Saltiel, M. Fichet, D. Bloch, M. Ducloy, Resonant coupling in the van der Waals interaction between an excited alkali atom and a dielectric surface: an experimental study via stepwise selective reflection spectroscopy. Eur. Phys. J. D **23**, 237 (2003)
11. H. Failache, S. Saltiel, M. Fichet, D. Bloch, M. Ducloy, Resonant van der Waals repulsion between excited Cs atoms and sapphire surface. Phys. Rev. Lett. **83**, 5467 (1999)
12. M. Oria, M. Chevrollier, D. Bloch, M. Fichet, M. Ducloy, Spectral observation of surface-induced van der Waals attraction on atomic vapour. Europhys. Lett. **14**, 527 (1991)
13. G. Dutier, S. Saltiel, D. Bloch, M. Ducloy, Revisiting optical spectroscopy in a thin vapor cell: mixing of renijéection and transmission as a Fabry Perot microcavity effect. J. Opt. Soc. Am. B **20**, 793 (2003)

Chapter 5
Atom–Atom Interactions

Up to this point, we have only alluded to the effect that the interactions between identical atoms have on the spectral properties. In many cases, these inter-particle interactions are negligible. However, if the atoms are separated by less than the emission wavelength, λ, resonant dipole–dipole interactions modify the radiative decay rate [1] and induce a splitting or shift of the resonance [2].

When these interactions become important, one can no longer consider the system as a sum of individual responses. Following the approach taken by Javanainen et al. [3], if correlations in the system are important, the interactions are said to be *cooperative*. If correlations are not important, which is the case when the system is subject to inhomogeneous broadening, the response is *collective*. In both cases the individual response is modified by its surroundings, but the two cases are subtly different.

This difference can be understood by considering the exchange of multiple photons between the dipoles. In the cooperative case, without inhomogeneous broadening, the photon exchange between any two emitters is always resonant and thus correlations are important. This is not the case when inhomogeneous broadening is present, however, as the dipoles can have different resonant frequencies. In this case the correlations are suppressed.

In this chapter we map out the transition between independent atoms and a fully interacting ensemble, moving smoothly between the two by increasing the density of atoms confined in the nano-cell. We can characterise the importance of the interactions in the system by the dimensionless parameter $\mathcal{C}$ (called, in various works, either the collectivity or cooperativity parameter), which is given by [4]

$$\mathcal{C} = 2\pi \, N k^{-3},\tag{5.1}$$

where, as in previous chapters, N is the atomic density and $k = 2\pi/\lambda$ with λ the transition wavelength. Physically this parameter reflects the number of atoms confined in a box whose dimensions are the reduced wavelength, $\lambda/2\pi = k^{-1}$, and hence interactions become important for $\mathcal{C} \geq 1$.

J. Keaveney, *Collective Atom–Light Interactions in Dense Atomic Vapours*,
Springer Theses, DOI: 10.1007/978-3-319-07100-8_5,

5.1 The Dipole–Dipole Interaction Between Two Identical Atoms

We start by looking at the interaction between two atoms. The interaction potential between two bodies can be expanded in a power series as [5]

$$V_{\mathrm{dd}}(r) = \frac{1}{\hbar} \sum_n \frac{C_n}{r^n} \tag{5.2}$$

where the C_n terms are coefficients, r is their separation and n is the order of the interaction. This is known as the multipole expansion, since the C_n/r^n each describe a particular interaction between charge distributions, starting with monopole–monopole terms and extending through monopole–dipole, dipole–dipole, dipole–quadrupole and so on. The lowest order terms are the longest ranged, and the physical bodies involved in the interaction (neutral atom, ion, molecule etc.) determine the lowest order term involved. For two identical neutral atoms, the lowest order interaction is for $n = 3$, the dipole–dipole interaction, and the interaction strength scales proportionally with the number density, since $r^{-3} \propto N$.

In general, the dipole–dipole interaction can be understood in terms of the pair states of the two contributing atoms. For two two-level atoms with ground state $|g\rangle$ and excited state $|e\rangle$, the bare atomic states and the pair states of the joint system are shown in panels (a) and (b) of Fig. 5.1, respectively. In the absence of dipole–dipole interactions, the energies of the singly excited states $|ge\rangle$ and $|eg\rangle$ are degenerate. The dipole–dipole interaction couples the two pair states, according to the interaction Hamiltonian

$$\mathcal{H}_{\mathrm{dd}} = \hbar \begin{pmatrix} 0 & V_{\mathrm{dd}} \\ V_{\mathrm{dd}} & \delta \end{pmatrix}, \tag{5.3}$$

where V_{dd} is the strength of the interaction, and δ is difference in energies of the initial and final pair states. When $\delta = 0$, the interaction is known as the *resonant dipole–dipole interaction*. For the case in Fig. 5.1, the interaction between the two singly excited states is resonant, since the initial and final pair states have the same energy. The new eigenstates of the coupled system are known as the Dicke states, $|\pm\rangle = \frac{1}{\sqrt{2}}(|eg\rangle \pm |ge\rangle)$, the symmetric and anti-symmetric superpositions of the two pair states, whose energies are shifted by the eigenvalues of the interaction Hamiltonian

$$\Delta E = \frac{\delta \pm \sqrt{\delta^2 + 4V_{\mathrm{dd}}^2}}{2}. \tag{5.4}$$

For resonant interactions, $\delta = 0$ and $\Delta E = \pm V_{\mathrm{dd}}$. When the two pair states do not have the same energy, there are two regimes of interest. At short range, $V_{\mathrm{dd}} \gg \delta$ and the eigenvalues approximate to $\pm V_{\mathrm{dd}}$, i.e. the interaction behaves in the same way as the resonant case. However at long range, known as the van der Waals regime where $V_{\mathrm{dd}} \ll \delta$, we find $\Delta E \approx \delta + V_{\mathrm{dd}}^2/\delta$, and $-V_{\mathrm{dd}}^2/\delta$. These non-resonant interactions are

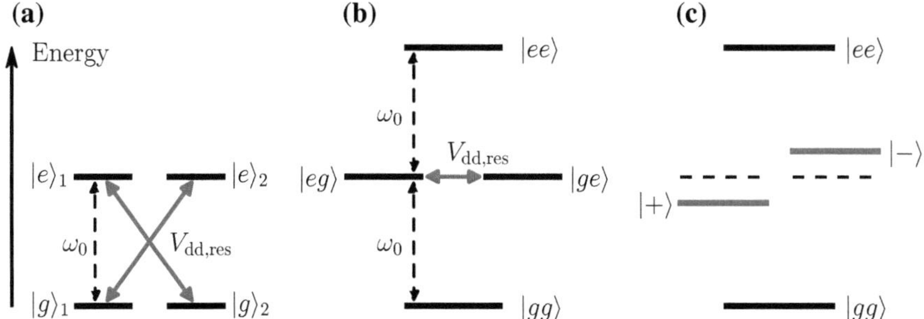

Fig. 5.1 Resonant dipole–dipole interaction for two two-level atoms in the individual and pair bases. **a** In the bare atomic basis, states $|g\rangle$ and $|e\rangle$ are separated by an energy $\hbar\omega_0$. **b,c** In the pair state picture, in the absence of dipole–dipole interactions the singly-excited states $|ge\rangle$ and $|eg\rangle$ are degenerate. The resonant dipole–dipole interaction lifts the degeneracies of the singly excited states, forming the Dicke states $|\pm\rangle = \frac{1}{\sqrt{2}}(|eg\rangle \pm |ge\rangle)$. Note that the energies of all the pair states will be shifted by induced (non-resonant) dipole–dipole interactions by a small amount, but for this system, these are small compared with the resonant interaction

Fig. 5.2 a Illustration of the dipole–dipole interaction between two atoms, showing the dipole moment vectors $\mu_{1,2}$ and the vector between them **r**, which creates an angle θ with the alignment axis (taken in the z-direction)

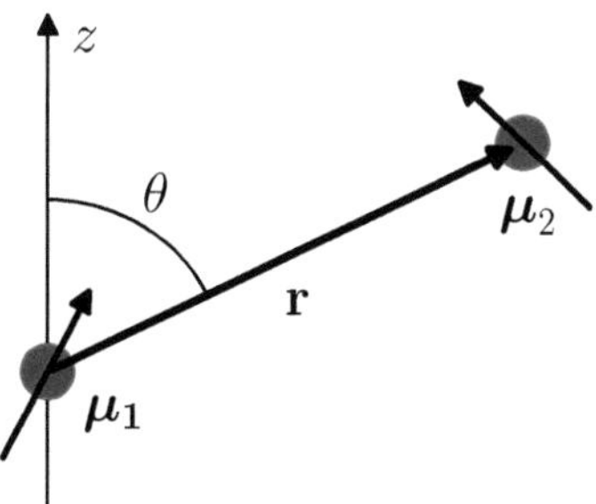

known as *induced* dipole–dipole interactions. Although there is always a contribution from these non-resonant interactions [6], they scale as $V_{\mathrm{dd}}^2 \propto C_6/r^6$ and therefore are unimportant when resonant interactions are dominant, which is the case in the present work. For further discussion on the dipole–dipole interaction see, for example, Ref. [6].

5.1.1 Static Dipoles

We first consider the case of two static dipoles, before considering the effect of oscillating dipoles. To avoid confusion, 'static' refers to the dipole moment vector, not the atomic motion (though this is also neglected for now). Figure 5.2 shows the geometry of the problem. Two dipoles with moment vectors μ_1 and μ_2 are displaced from each other by **r**, which forms an angle θ with the polarisation axis, in this case in the x-direction.

If we consider dipole 1 to be at the origin, then the second will experience an electric field given by [7]

$$\mathbf{E}_{\text{dipole}}(\mathbf{r}) = \frac{1}{4\pi\epsilon_0} \left(3\frac{\boldsymbol{\mu}_1 \cdot \mathbf{r}}{r^5}\mathbf{r} - \frac{\boldsymbol{\mu}_1}{r^3} \right). \tag{5.5}$$

The second atom will then experience an interaction with the dipolar field due to its own dipole moment

$$\hbar V_{\text{dd}} = -\boldsymbol{\mu}_2 \cdot \mathbf{E}_{\text{dipole}}. \tag{5.6}$$

Combining the two, we obtain

$$V_{\text{dd}} = \frac{1}{4\pi\epsilon_0\hbar} \left(\frac{\boldsymbol{\mu}_1 \cdot \boldsymbol{\mu}_2}{r^3} - 3\frac{(\boldsymbol{\mu}_1 \cdot \mathbf{r})(\boldsymbol{\mu}_2 \cdot \mathbf{r})}{r^5} \right). \tag{5.7}$$

In the presence of an external field, the dipoles will align themselves to the polarisation of the field. For the case presented in Fig. 5.2, the dipoles will align so that $\boldsymbol{\mu}_1 = |\boldsymbol{\mu}_1|\hat{\mathbf{z}}$ and $\boldsymbol{\mu}_2 = |\boldsymbol{\mu}_2|\hat{\mathbf{z}}$. If $|\boldsymbol{\mu}_1| = |\boldsymbol{\mu}_2| = \mu$, the interaction simplifies to

$$V_{\text{dd}} = -\frac{\mu^2}{4\pi\epsilon_0\hbar} \left(\frac{3\cos^2(\theta) - 1}{r^3} \right). \tag{5.8}$$

The sign of V_{dd} therefore changes depending on the arrangement of the dipoles. This is similar to the case of two bar magnets, aligned along the vectors $\boldsymbol{\mu}_{1,2}$. When the magnets are end-to-end ($\theta = 0$) opposite poles are near each other and the interaction is attractive ($V_{\text{dd}} < 0$) and therefore leads to a red-shift of the resonance, whereas if they are placed side-to-side ($\theta = \pi/2$), the interaction is repulsive with a corresponding blue-shift.

For an isotropic distribution of two-atom pairs, taking the angular average over the 4π solid angle gives $\langle V_{\text{dd}}(\theta)\rangle = 0$. For this interaction to be non-zero, one must also consider the effect of the propagation phase due to the laser. However, in our experiment we have a pseudo-2D system because of the confinement in the propagation direction. Taking the limit of this, the angular average reduces to an average over θ and hence $\langle\cos^2(\theta)\rangle = 0.5$ implying a non-zero dipole–dipole interaction regardless of the propagation phase. Clearly, this reduced dimensionality will depend on the thickness of the medium, and one therefore expects a geometry dependent interaction. This will be discussed in more detail later in this chapter.

To quantitatively determine when the dipole–dipole interaction becomes significant, we can set $V(r)$ equal to the natural linewidth Γ_0, and rearrange. We can express μ^2 as [8]

$$\mu^2 = \frac{3\pi\epsilon_0\hbar}{k^3}\frac{(2J_e + 1)}{(2J_g + 1)}\Gamma_0, \tag{5.9}$$

with

$$\mu = \langle J_g | er | J_e \rangle. \tag{5.10}$$

Then the dipole–dipole interaction becomes significant when

$$\frac{3}{8} \frac{1}{(kr)^3} \frac{(2J_e + 1)}{(2J_g + 1)} = 1, \tag{5.11}$$

which is satisfied when

$$kr = \sqrt[3]{\frac{3(2J_e + 1)}{8(2J_g + 1)}}. \tag{5.12}$$

For the D1 line, this yields $kr = \sqrt[3]{3/8} \simeq 0.7$, while for the D2 line, we find $kr = \sqrt[3]{3/4} \simeq 0.9$; or in terms of the collectivity parameter, $C \sim 2\pi$.

5.1.2 Oscillating Dipoles

For oscillating dipoles, there is a relative phase of e^{ikr} between the dipoles which is due to the finite propagation speed of light. The full electric field of Eq. (5.5) then has the more general form [9, 10]

$$\mathbf{E}_{\text{dipole}}(\mathbf{r}) = \frac{1}{4\pi\epsilon_0} \left[\left(\frac{1}{r^3} - \frac{ik}{r^2} \right) (3\hat{\mathbf{r}}(\hat{\mathbf{r}} \cdot \boldsymbol{\mu}_1) - \boldsymbol{\mu}_1) + \frac{k^2}{r} (\hat{\mathbf{r}} \times \boldsymbol{\mu}_1) \times \hat{\mathbf{r}} \right] e^{ikr}, \tag{5.13}$$

where $r = |\mathbf{r}|$ and $\hat{\mathbf{r}} = \mathbf{r}/r$. The static case can be recovered by setting $k = 0$, i.e. an oscillation with infinite wavelength. As before, we assume that an incident electric field polarises the atoms so that their dipole moments align. In terms of Γ_0, the complete pair potential has the form

$$V_{\text{dd}} = -\frac{3\Gamma_0}{4} \left[\frac{1 - ikr}{(kr)^3} (3\cos^2\theta - 1) + \frac{1}{kr} \sin^2\theta \right] e^{ikr}, \tag{5.14}$$

In contrast to the static case of Eq. (5.7), the dipole–dipole interaction with oscillating dipoles is a complex quantity. The real part, as in the static case, modifies the energy levels, while the imaginary part modifies the radiative decay rates. The new decay rates and eigenenergies are given by

$$\Gamma_\pm = \Gamma_0 \pm \Gamma_{\text{dd}}, \tag{5.15a}$$

$$\Delta_\pm = \Delta \pm \Delta_{\text{dd}}, \tag{5.15b}$$

which are related to the real and imaginary parts of V_{dd} by [11]

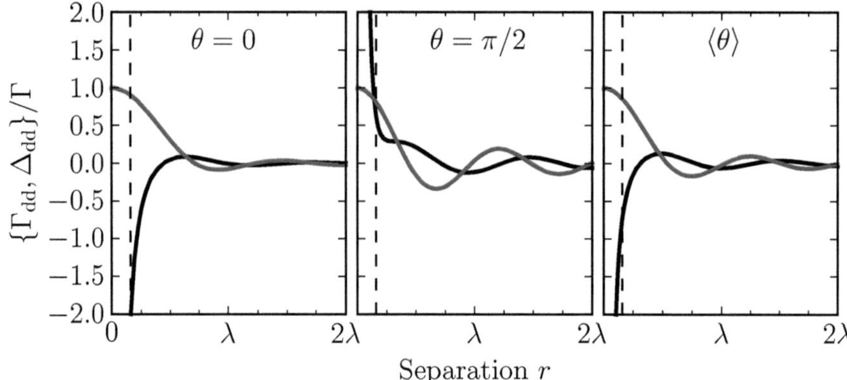

Fig. 5.3 Dipole–dipole induced shift Δ_{dd} (*black*) and width Γ_{dd} (*red*) as a function of separation, for $\theta = 0$, $\theta = \pi/2$ and the angular average $\langle\theta\rangle$ (where $\langle\sin^2(\theta)\rangle = \langle\cos^2(\theta)\rangle = 1/2$), based on Eq. (5.14). The *dashed vertical lines* are at $kr = 1$

$$\Gamma_{dd} = -2\,\mathrm{Im}(V_{dd}), \tag{5.16a}$$

$$\Delta_{dd} = \mathrm{Re}(V_{dd}), \tag{5.16b}$$

where the $\pm$ subscripts refer to the Dicke states $|+\rangle$ and $|-\rangle$. Figure 5.3 shows how the interaction scales with the separation of the dipoles, for the two cases $\theta = 0$ and $\theta = \pi/2$ which correspond to the maximal attractive and repulsive interactions in the static case, respectively. Also plotted is the angular average, $\langle\theta\rangle$, which has a net red-shift at small separations. The dotted line in each panel is at $kr = 1$. For separations greater than this, we are in the far field and only the long-ranged $1/r$ term remains, which combined with the e^{ikr} term make up a decaying spherical wave. The amplitude of this is much smaller than Γ_0, so the dipole–dipole interaction between any two atoms in this regime is negligible. However, this term can still be important for an ensemble of dipoles, as will be discussed later in this chapter. In the limit of zero separation, we find that $\Gamma_{dd} \to \Gamma_0$. We see that the decay rate of the Dicke states $|+\rangle$ and $|-\rangle$ are now enhanced and suppressed, respectively. State $|+\rangle$ decays twice as fast as the natural rate, a process known as superradiance, whilst the antisymmetric state does not decay at all, called subradiance [12]. As the separation is varied, the decay rate oscillates between super- and sub-radiant, first observed by DeVoe and Brewer in a landmark experiment with two trapped ions [1].

For separations $kr \leq 1$ the near field is dominated by the $1/r^3$ term, and hence the shift of the lines diverges - either to lower or higher energies, depending on orientation, in agreement with the static case.

Figure 5.4 shows the angular dependence of Γ_{dd} and Δ_{dd}, for a fixed separation of $kr = 1$. In this case the real part gives rise to a shift that oscillates with θ, in the same way as the static case. The decay rate modification due to the imaginary part of V_{dd} also varies with the orientation, but much less significantly than the real part.

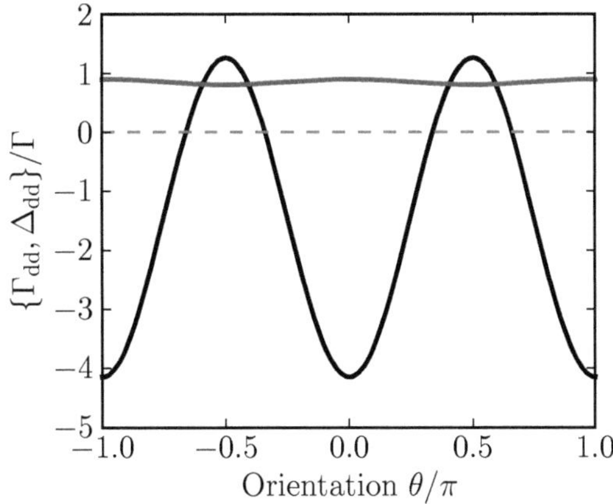

Fig. 5.4 Dipole–dipole induced shift Δ_{dd} (*black*) and width Γ_{dd} (*red*) as a function of the dipole orientation angle θ, for a separation $kr = 1$, based on Eq. (5.14)

5.1.3 Potential Curves

To illustrate the effect of the dipole–dipole interaction for a complex atomic system, we calculate the energies of the pair states directly. Here the interaction is only included for the static case, with $\theta = 0$. The interaction potential is then simply

$$V_{dd}(r) = \frac{1}{2\pi\epsilon_0\hbar}\frac{\mu^2}{r^3}.$$ (5.17)

Before calculating this interaction for the full atomic system, it is useful to first look at the simplest case, that of a pair of two-level atoms, as shown in Fig. 5.1. In the absence of interactions, the Hamiltonian for this system using the pair state basis $\{|gg\rangle, |ge\rangle, |eg\rangle, |ee\rangle\}$ is

$$\mathcal{H}_{\text{pair}} = \hbar \begin{pmatrix} 0 & 0 & 0 & 0 \\ 0 & \omega_0 & 0 & 0 \\ 0 & 0 & \omega_0 & 0 \\ 0 & 0 & 0 & 2\omega_0 \end{pmatrix}.$$ (5.18)

Neglecting the off-resonant interactions, the resonant dipole–dipole interaction then couples the two singly-excited states

$$\mathcal{H}_{dd} = \hbar \begin{pmatrix} 0 & 0 & 0 & 0 \\ 0 & 0 & V_{dd} & 0 \\ 0 & V_{dd} & 0 & 0 \\ 0 & 0 & 0 & 0 \end{pmatrix}.$$ (5.19)

The total matrix $\mathcal{H}_{\text{pair}} + \mathcal{H}_{dd}$ can then be diagonalised to find the eigenvalues of the coupled interacting system.

For a real atomic system, there are many more states, coupled by $\sigma^{\pm}$ and π transitions according to the dipole selection rules [13]. We can write out the full atomic structure using the uncoupled basis in terms of the quantum numbers L, S and I, (the orbital, spin and nuclear angular momenta, respectively). In this basis, any state of the system can be written as $|L, m_L, S, m_S, I, m_I\rangle$, and the Hamiltonian for the S ($L = 0$) to P ($L = 1$) transition is a square matrix with dimensions $4(2S+1)(2I+1)$. This uncoupled Hamiltonian has previously been used extensively to predict transitions in Rb [8, 14] and Cs [15], particularly in the presence of applied magnetic fields [16–18]. Ignoring the substructure, we can write the Hamiltonian as a 4×4-matrix; the atomic part can be written as

$$
\mathcal{H}_{\text{atom}} = \begin{pmatrix} \mathcal{H}_S & 0\,0\,0 \\ 0 \\ 0 & \begin{pmatrix} \ddots & \\ & \mathcal{H}_P \\ & & \ddots \end{pmatrix} \\ 0 \end{pmatrix} . \tag{5.20}
$$

where each element inside the matrix is itself another matrix of size $(2L+1) \times (2S+1) \times (2I+1)$, and the P-state is broken down into terms with $\Delta m_L = 0, \pm 1$. The two-atom pair Hamiltonian is then expressed as

$$
\mathcal{H}_{\text{pair}} = \mathbb{1} \otimes \mathcal{H}_{\text{atom}} + \mathcal{H}_{\text{atom}} \otimes \mathbb{1}, \tag{5.21}
$$

where the $\otimes$ symbol represents the tensor product and $\mathbb{1}$ is an identity matrix of the same dimensions as $\mathcal{H}_{\text{atom}}$. The dipole–dipole interaction becomes more complex in this basis, as we need to take into account coupling states with different angular momenta ($\sigma^{\pm}$ and π transitions). In the static case, when the dipoles align with the applied field (and each other) and for $\theta = 0$, the interaction potential can be written as [19]

$$
V_{\text{dd}} = \frac{\mu_{1+}\mu_{2-} + \mu_{1-}\mu_{2+} - 2\mu_z\mu_z}{4\pi\epsilon_0 \hbar r^3}, \tag{5.22}
$$

where the $+, -, z$ are associated with the σ^+, σ^-, π transitions, respectively. In a two-atom basis, we write this as

$$
V_{\text{dd,pair}} = \frac{1}{4\pi\epsilon_0 \hbar r^3} \left(\mu_+ \otimes \mu_- + \mu_- \otimes \mu_+ - 2\mu_z \otimes \mu_z \right), \tag{5.23}
$$

where the $\mu_{+,-,z}$ are the matrices

$$
\mu_+ = \mu \begin{pmatrix} 0 & 0 & 0 & 1 \\ 0 & 0 & 0 & 0 \\ 0 & 0 & 0 & 0 \\ 1 & 0 & 0 & 0 \end{pmatrix}, \quad \mu_- = \mu \begin{pmatrix} 0 & 1 & 0 & 0 \\ 1 & 0 & 0 & 0 \\ 0 & 0 & 0 & 0 \\ 0 & 0 & 0 & 0 \end{pmatrix}, \quad \mu_z = \mu \begin{pmatrix} 0 & 0 & 1 & 0 \\ 0 & 0 & 0 & 0 \\ 1 & 0 & 0 & 0 \\ 0 & 0 & 0 & 0 \end{pmatrix}, \tag{5.24}
$$

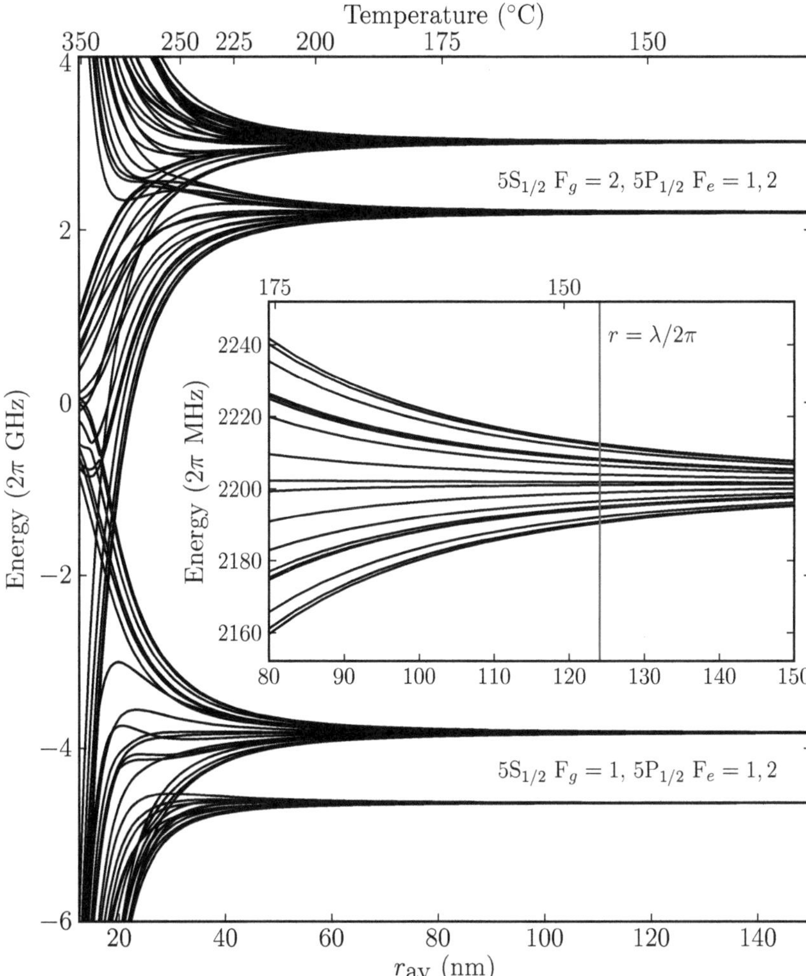

Fig. 5.5 Pair-state energies in the two-atom basis for the $5S_{1/2}$–$5P_{1/2}$ transition in ^{87}Rb. The main plot shows the energies of the $5S_{1/2}$, $5P_{1/2}$ pair state manifolds, with both the ground and excited state hyperfine structure. As the atomic separation is decreased, the degeneracies of the states are lifted, resulting in a splitting and shift of the lines at small separations. The inset shows a zoom of the $5S_{1/2}$ F$_g$ = 2, $5P_{1/2}$ F$_e$ = 1 manifold, where the states start to split. At a separation $r = \lambda/2\pi$ the maximum splitting is $\pm 3\Gamma_0/2$, in line with Eq. (5.17). Zero on the energy scale represents the line centre energy of the ^{87}Rb $5S_{1/2}$, $5P_{1/2}$ pair state without hyperfine structure at infinite separation

and the $\mathbb{1}$'s are identity matrices of size $(2S + 1)(2I + 1)$. For the Rb ground state transition we have $L = 0$ or 1, $S = 1/2$ and $I = 3/2$ or $5/2$ (for ^{87}Rb or ^{85}Rb, respectively). The total matrix $\mathcal{H}_{\mathrm{pair}} + V_{\mathrm{dd,pair}}$ is then diagonalised to find the eigenenergies of the pair states.

Figure 5.5 shows the energy level structure as a function of the average interatomic spacing, $r_{av} = 5/9N^{-1/3}$. We plot to a minimum separation $r_{av} \approx r_W$, the limit of the binary approximation (see Sect. 5.2.2). For computational convenience, we have taken the case of ^{87}Rb ($I = 3/2$ for ^{87}Rb as compared to $I = 5/2$ for ^{85}Rb, making the matrix smaller and hence computationally less demanding). The molecular states are broadly split into SS, SP+PS, and PP. The P-states can be broken down further into their fine-structure components, $P_{1/2}$ and $P_{3/2}$, split by $\sim$7 THz. We show in the figure the lines in the $5S_{1/2}$, $5P_{1/2}$ manifold. At large separations the next biggest splitting is then the S-state hyperfine interaction (6.8 GHz), and finally the P-state hyperfine levels (800 MHz). The Zeeman sublevels are included in the Hamiltonian but are degenerate in the absence of an applied magnetic field.

As the atomic separation decreases, the lines that are degenerate at infinite separation diverge, which becomes more significant as the separation decreases. For relatively large separations the lines split symmetrically (inset), and at a separation $r = \lambda/2\pi$, the maximum splitting is $\pm 3\Gamma_0/2$, as expected from Eq. (5.17). As the separation decreases further, some lines turn over indicating an overall shift of the line.

5.2 Dipole–Dipole Interaction for an Ensemble of Atoms

The total dipole–dipole interaction energy that one atom experiences can be described as a sum over all of the other atoms in the system. For an isotropic medium with density N, the total interaction is given by an integral over the whole space

$$V_{\text{tot}} = N \int_0^\infty 4\pi r^2 \, V(r) \, \mathrm{d}r. \tag{5.25}$$

For the resonant dipole–dipole interaction of Eq. (5.14) there are terms proportional to r^{-3}, r^{-2} and r^{-1}, hence even though the furthest atoms contribute least, there are more of them so their sum still remains important, not just the nearest neighbour. The problem with Eq. (5.25) is because of the r^{-3} term, the integral diverges at $r = 0$. The usual approach (see Ref. [10, 20, 21]) is to set the lower limit to some value a, which essentially places the dipole in a spherical cavity of radius a. This dipole is then subject to a local electric field $\boldsymbol{E}_{\text{loc}}$, which is known as the Lorentz local field [22],

$$\boldsymbol{E}_{\text{loc}} = \boldsymbol{E} + \frac{\boldsymbol{P}}{3\epsilon_0}, \tag{5.26}$$

where $\boldsymbol{E}$ is the applied field and $\boldsymbol{P} = N\langle\boldsymbol{\mu}\rangle$ is the macroscopic polarisation of the medium, related to the average individual dipole moment. The underlying mechanism of light scattering is the interference between the incident field and the local field produced by induced oscillatory dipoles, as in the Ewald-Oseen extinction theorem [23]. In a dense medium, this dipolar field produced by the surrounding dipoles modifies the optical response of each individual dipole.

5.2.1 Lorentz Shift

By relating the macroscopic to the microscopic response of the system, we can calculate the effect of the dipole–dipole interaction across the medium. The susceptibility χ determines the macroscopic response of the medium, whereas the polarisability α determines the individual atomic response. In terms of the polarisability

$$P = 4\pi\epsilon_0 N\alpha E_{\mathrm{loc}} = 4\pi N\alpha(\epsilon_0 E + P/3) \tag{5.27}$$

and solving for P we find a relation between the macroscopic variable χ and the single dipole parameter α which is referred to as the Lorentz–Lorenz law [22]

$$\chi = \frac{4\pi N\alpha}{1 - \frac{4}{3}\pi N\alpha}. \tag{5.28}$$

The polarisability is related to the weak-probe susceptibility for independent atoms (Eq. 2.22) by

$$\alpha = \frac{\chi_{\mathrm{atom}}}{4\pi N}, \tag{5.29}$$

and substituting for α we find

$$\chi = -\frac{N\mu^2/\epsilon_0\hbar}{\Delta + i\Gamma_0/2 + N\mu^2/3\epsilon_0\hbar}, \tag{5.30}$$

which is equivalent to Eq. (2.22) but with a density-dependent shift in the resonance frequency known as the *Lorentz shift*

$$\Delta_{\mathrm{LL}} = -\frac{N\mu^2}{3\epsilon_0\hbar}. \tag{5.31}$$

5.2.2 Collisional Effects

In a thermal ensemble the medium is never really isotropic—at any given time the atoms are randomly distributed. For a random distribution of atoms, the spacing between atoms r has the distribution [24]

$$W(r) = 4\pi Nr^2 e^{-4\pi/3Nr^3}, \tag{5.32}$$

whose average value is

$$r_{av} = \int_0^\infty r\,W(r)\,\mathrm{d}r \approx \frac{5}{9}\,N^{-1/3}, \qquad (5.33)$$

and in a collision the atoms can get much closer than this. These collisional processes have been previously studied in detail (see, for example, Foley [25], Lewis [6] or Thorne [5]). We assume that any collision involves only two atoms, known as the *binary approximation*, which is valid as long as [5]

$$\frac{4\pi}{3}Nr_{\mathrm{W}}^3 < 1, \qquad (5.34)$$

where r_{W} is known as the Weisskopf radius. This is a statement that if an atom is at the centre of a sphere, there will be at most one other atom within a distance r_{W}. The radius depends on the interaction strength, which is determined by μ, and the mean relative velocity of the atoms, given by [26]

$$\bar{v} = \sqrt{16k_{\mathrm{B}}T/\pi m}. \qquad (5.35)$$

The interaction strength can be characterised by the coefficient β (known as the self-broadening coefficient, discussed below), and the Weisskopf radius is then [26]

$$r_{\mathrm{W}} = \sqrt{\frac{\beta}{2\pi\bar{v}}}. \qquad (5.36)$$

For the D2 line in Rb, $r_{\mathrm{W}} = 14$ nm, so the binary approximation breaks down at a density $N_{\mathrm{binary}} \approx 10^{17}$ cm^{-3}. In a thermal vapour, this corresponds to a temperature $T \approx 360\,^\circ$C.

Beyond this density, multi-perturber interactions must be considered [27–29], which are beyond the scope of this thesis. There are two widely discussed cases of collisional effects, which are valid for different parameter regimes; the impact approximation and the quasistatic approximation.

The impact approximation assumes the collisions are described by a classical trajectory with an impact parameter that represents the closest approach distance. The strength of the dipole–dipole interaction is therefore transient, causing a dephasing of the dipoles. In the simplest (Lorentz) model, this process is described by a decay process with a lifetime, and associated linewidth $\tau_{\mathrm{impact}} = 1/\Gamma_{\mathrm{impact}}$. This approximation assumes that an atom moves a significant amount during a collision, which is valid at low density with fast-moving atoms for resonant (or near-resonant) excitation [5].

The quasistatic approximation assumes the opposite conditions to the impact approximation; the atom does not move significantly during the collision, and so the motion does not play a significant role; in this case the interaction reduces to the case presented in the previous section. This approximation is valid for high density vapour, where the atomic motion is slow, and for off-resonant excitation.

Both of these approximations, however, give essentially the same results - an increase in the linewidth and a shift of the line centre that are both linearly proportional to the density. In the binary approximation, the collisional linewidth (FWHM) is given by [14]

$$\Gamma_{\text{col}} = \beta N = \frac{N\mu^2}{3\hbar\epsilon_0}\sqrt{\frac{2J_g+1}{2J_e+1}} = 2\pi\sqrt{\frac{2J_e+1}{2J_g+1}}\,\Gamma_0 N k^{-3}, \qquad (5.37)$$

or in terms of the collectivity parameter

$$\Gamma_{\text{col}} = \sqrt{\frac{2J_e+1}{2J_g+1}}\,\Gamma_0 \mathcal{C}, \qquad (5.38)$$

For the D2 line, the self-broadening coefficient $\beta = 2\pi \times 1.032 \times 10^{-7}\,\text{Hz}\,\text{cm}^3$ [14]. The collisional linewidth has a Lorentzian profile, which must be convolved with the Voigt profile of Eq. (2.28). However, we can take advantage of the associative property of convolution and the fact that a convolution of two Lorentzians yields another Lorentzian whose width is the sum of the two, so the total Lorentzian component of the linewidth is just

$$\Gamma_{\text{tot}} = \Gamma_0 + \Gamma_{\text{col}}. \qquad (5.39)$$

This simplifies the analysis so that including collisional broadening into the weak probe susceptibility model we need only make the substitution $\Gamma_0 \to \Gamma_{\text{tot}}$.

The collisional shift is less well documented and has no simple analytic form. Friedberg et al. [21] suggest that the collisional shift should be a blue-shift. Experimental data is limited, as in linear optics, separating out the collisional shift from the Lorentz shift is challenging, since they are both linearly dependent on density and are the same order of magnitude [21].

However, Maki et al. [30] showed the collisional shift can be separated from the Lorentz shift in a nonlinear optical process, finding empirically that on the potassium D lines, $\Delta_{\text{col}} \approx \Delta_{\text{LL}}/1.4$ for the D1 line and $\Delta_{\text{col}} \approx \Delta_{\text{LL}}/4.8$ for the D2 line, i.e. the collisional shift is a red-shift.

5.3 Collective Effects

One must also consider the effect of collective interactions. In a thermal gas, these effects are usually unimportant, but when the density becomes high enough the interactions can dominate completely, as we shall demonstrate experimentally later in the chapter. The timescale due to the dipole–dipole interaction in this regime can be much smaller than any timescale due to motion, so a quasistatic picture is recovered and we once again treat the medium as an ensemble of motionless dipoles.

5.3.1 Cooperative Lamb Shift

The ability to measure the cooperative Lamb shift (CLS) is one such consequence of this quasistatic picture. This shift is also known as the collective Lamb shift, see for example Refs. [31, 32], but in this work we retain its historic name. It is named after the Lamb shift [33] whereby the interaction with the vacuum field induces a shift (radiative correction) in the energy levels. This interaction can be viewed as the electron emitting and subsequently reabsorbing a virtual photon [21].

The CLS is an extension of this picture to two atoms—the emission and absorption of virtual photons now occurs between pairs of atoms [21]. This leads to an additional energy shift of the transition frequencies that are dipole-coupled. Crucially, this is a coherent process and the net observable shift depends on the geometry of the problem.

Qualitatively, the effect of the CLS can be understood as the anisotropy of the dipole–dipole interaction combining with the anisotropy of the spatial distribution of atoms in the thin cell. Consider an atomic vapour with macroscopic thickness, and take an individual atom within the vapour. The dipole–dipole interaction this atom experiences depends in principle on every other atom in the sample, but for the sake of this argument, consider the dipole–dipole interaction to have a finite range, which we take to be R. Assuming we are far from the walls, the atom experiences interactions within a sphere, centred on itself. The total dipole–dipole interaction is then a sum of each pairwise interaction, with separation r and angle θ, as discussed earlier in the chapter. Assuming uniform atomic density, θ is also uniformly distributed between the first atom and any other neighbouring atom. The total interaction will then be a red-shift, owing to the anisotropy of the dipole–dipole interaction (see Fig. 5.4). However, as the thickness of the medium decreases below the radius of the sphere, the angular distribution becomes skewed, with θ more likely to be close to 0 or π, leading to a greater red-shift of the lines for the thinnest vapours. For oscillating dipoles the phase factor e^{ikr} complicates this simple picture, and if the dipoles are aligned by the presence of an external laser field (propagating in the z-direction) then there is an additional phase factor $e^{i\kappa(z-z')}$, where κ is the wavevector of the laser field and the two atoms are at positions z and z'.

In the limit of many atoms, the sum becomes an integral over the coordinate space. For a slab geometry, the excitation region forms a cylinder with radius b much bigger than its thickness ℓ (a pancake shape). With the atomic dipoles aligned with the laser polarisation, the total shift due to the interaction is the treble integral [21]

$$\Delta_{CLS} = \lim_{a \to 0} -\frac{N\mu^2}{\hbar\epsilon_0 \ell} \, \mathrm{Re} \int_0^\ell dz \int_0^\ell dz' \int_a^\infty 2\pi b \, db \, e^{i\kappa(z-z')} V_{\mathrm{dd,cyl}}(b, z - z'), \quad (5.40)$$

with the interaction potential in cylindrical coordinates given by

$$V_{\mathrm{dd,cyl}}(b, z - z') = \left[\frac{1}{2}\left(1 + \frac{(z - z')^2}{r^2}\right)\frac{k^2}{r}\right.$$
$$\left. + \frac{1}{2}\left(3\frac{(z - z')^2}{r^2} - 1\right)\left(\frac{ik}{r^2} - \frac{1}{r^3}\right)\right] e^{ikr}. \qquad (5.41)$$

The integrals in (5.40) are over the first atom at position z, and the second atom a distance r away, where $r^2 = b^2 + (z - z')^2$, and for a near-resonant laser field, $\kappa \approx k$.

For the case presented in this work where $b \gg \ell$, the solution of this integral produces a total shift that depends on the density of the medium and the thickness [21]

$$\Delta_{\mathrm{dd}} = \Delta_{\mathrm{LL}} + \frac{3}{4}|\Delta_{\mathrm{LL}}|\left(1 - \frac{\sin 2k\ell}{2k\ell}\right). \qquad (5.42)$$

The first term is the familiar Lorentz shift and the second term is the CLS, which depends on twice the propagation phase $k\ell$ due to the re-radiation by the second dipole. In a thermal vapour, the total shift is then the shift of Eq. (5.42) plus the collisional shift

$$\Delta_{\mathrm{tot}} = \Delta_{\mathrm{dd}} + \Delta_{\mathrm{col}}. \qquad (5.43)$$

There are two remarkable features of Eq. (5.42). First, it is only in the limit of zero thickness that the Lorentz shift can be measured; in the limit of a thick slab with $k\ell \gg 1$ then one would measure only a quarter of the Lorentz shift. The second feature is that the CLS is a shift to higher energy, which can be understood from the pairwise potential; in a thin slab all the dipoles lie in the plane, and therefore oscillate in phase, such that the dipole–dipole interaction reduces to the static case. After averaging over θ this gives the maximal attractive interaction resulting in the Lorentz shift to lower energy.

As one moves out of the plane in the propagation direction the relative phase of the dipoles changes, allowing more interactions with a blue-shift in the angular average.

Though the CLS has been the topic of many theoretical works, including studies of the CLS in various geometries [34, 35] and the link between superradiance and CLS [31, 36–38], experimental observations of the CLS are few in number, limited to two examples (other than the present work) involving three-photon excitation in the limit of the thickness being much larger than the transition wavelength in an atomic gas [39], and x-ray scattering from Fe layers in a planar cavity [32, 40].

5.3.2 Cooperative Decay Rate and Superradiance

With two atoms, we showed in the previous section that the real part of the dipole–dipole interaction gives rise to a shift of the lines, whilst the imaginary part is responsible for a modification of the decay rate. It is therefore logical that this fundamental link between dipole–dipole induced shift and decay rate also extends to systems involving a large number $\mathcal{N}$ of atoms.

The enhancement or suppression of the decay rate, super- and sub-radiance, in atomic ensembles is a complex problem that has been extensively studied theoretically [41], with many experimental verifications of superradiant emission in various systems [42–44].

Sub-radiance, on the other hand, has only been observed for two atoms [1]; the observation of $\mathcal{N}$-atom subradiance remains a current area of interest [45, and refs. therein].

Superradiance arises because correlations between atoms in an ensemble cause the atoms to emit coherently and in phase with one another [41]. The emitted field is proportional to $\mathcal{N}$ and therefore the intensity proportional to $\mathcal{N}^2$. By an energy conservation argument, the duration of this emission must be shorter than for $\mathcal{N}$ independent emitters, and hence the peak rate of this emission must be higher. The emission process itself is normally considered as an ensemble of spin-1/2 particles with spin up and down representing the excited and ground-states of the atom, prepared in the symmetric state with all spins up.

In the time domain, the decay of this system leads to a characteristic emission curve (Fig. 1b of Ref. [41]) as the decay rate reaches a maximum value proportional to $\mathcal{N}^2$ when half of the atoms have decayed and then drops to zero at late time [12].

In a similar way to the cooperative Lamb shift, the additional phase factor due to the laser field also affects the superradiant emission, causing it to be strongly peaked in the propagation direction [12, 46].

For the two-atom system, the real part of the dipole–dipole interaction is responsible for the frequency shift, whilst the imaginary part is responsible for super- and sub-radiance. For the $\mathcal{N}$-atom case, a similar argument holds. In the slab geometry, the real part gives rise to the CLS. The corresponding modification to the decay rate is known as the cooperative decay rate, and has been the subject of several recent theoretical works [35, 47]. However, as the decay rate depends on the degree of excitation in the medium [12], it is unlikely to be a large factor to this experiment, where only a weak probe is used. This is in contrast to the CLS, which can be observed in the limit of weak excitation where there is negligible population of the excited state. Indeed, the magnitude of the shift decreases with excitation of the medium, cancelling completely when half the population is in the excited state [21], in a similar way to the collisional linewidths investigated in Ref. [28, 48, 49].

5.3.3 A Note on Dipole Blockade

Dipole blockade is an effect whereby the interactions between atoms prevent multiple excitation in a medium, first proposed by Lukin et al. [50]. To illustrate the effect, let us consider the two-atom, two-level picture as in Fig. 5.1, but with only the long range induced (non-resonant) dipole–dipole interactions, $V_{dd}(r) \propto r^{-6}$. The strength of the induced dipole–dipole interactions is strongly dependent on the (effective) principal quantum number, scaling as n^{*11}, so this effect is most relevant to Rydberg atomic systems.

When the atoms are far apart from each other V_{dd} is smaller than the laser linewidth γ_{laser} and so the doubly-excited state $|ee\rangle$ can be populated. However, as the separation decreases V_{dd} becomes significant, and when $V_{dd} > \gamma_{laser}$ then $|ee\rangle$ can no longer be populated. [1]

Thus, around a Rydberg atom there exists a 'blockade radius' where a second atom cannot be excited. Using this mechanism it is possible to construct a quantum gate [51]—the state of a 'control' atom, either in the Rydberg state or not, affects what happens to a second 'target' atom. Rydberg atoms are therefore an attractive candidates for constructing quantum bits (qubit) in quantum information processing (QIP) applications.

An interesting question is therefore whether it is possible to have a similar blockade mechanism in a simpler system, on a single ground-state transition. Returning to Fig. 5.1, but now considering the resonant interaction, the effect of the interaction is to split the singly-excited states. Immediately then one can conclude that any blockade effect will be of completely different character; all resonant excitation will be blockaded, given a suitable interaction strength. However, off-resonant excitation (with respect to the unshifted energy levels) could still exhibit blockade-type behaviour, since only the singly-excited states are shifted.

For an ensemble of thermal atoms at high densities, the resonant dipole–dipole interaction can be the dominant process, over an order of magnitude larger than the thermal effects (Doppler broadening). As has already been discussed, in this regime the effect of motional dephasing can once again be neglected. However, the increase of the linewidth due to collisional broadening plays an important role in the blockade mechanism. Comparing the line broadening to the shift, which we can see from examination of Eqs. (5.37) and (5.42) we find (for the D2 line) $\Delta_{LL} = \Gamma_{tot}/\sqrt{2}$. Assuming the collisional shift is small compared to the Lorentz shift, then we conclude that in a thermal vapour the shift due to resonant dipole–dipole interactions is always smaller than the increase in linewidth [52]. Hence there can be no blockade effect in thermal atoms, since the excitation laser frequency will always be in the wings of the resonance.

5.4 Experimental Results

We use transmission spectroscopy to investigate the modification to the linewidth and shifts due to dipole–dipole interactions. The experimental setup and calibration of the signals are described in appendix B.

---

[1] The excited state linewidth for Rydberg atoms is very much narrower than for low lying excited states so that it is usually only the laser linewidth that is the limiting factor in the blockade mechanism.

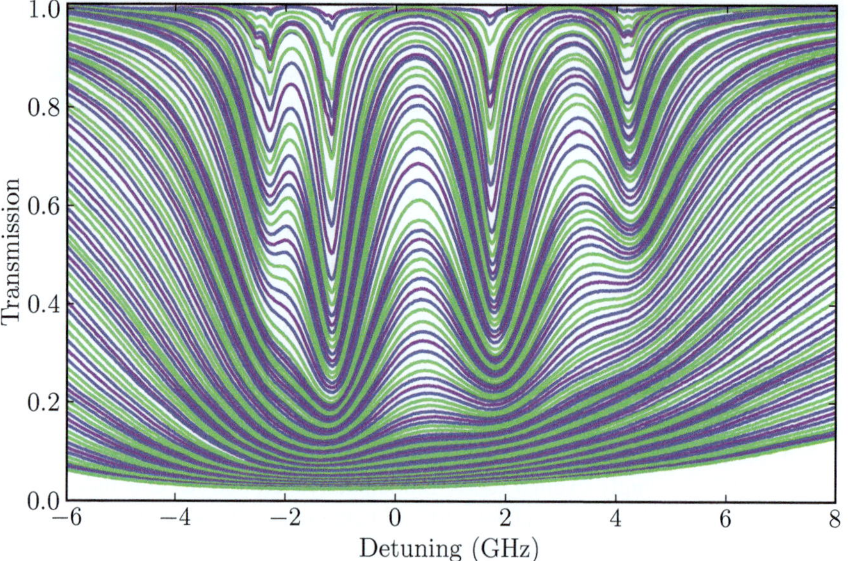

Fig. 5.6 Evolution of the transmission spectrum with increasing atomic density as a function of linear detuning $\Delta/2\pi$. Experimental data showing $\sim$100 individual traces with vapour thickness $\ell = \lambda/2$ as temperature is smoothly varied from 120 °C (*topmost line*) to 330 °C (*bottom*). For low atomic density, the optical depth is small and interactions between atoms are negligible, allowing the Dicke narrowed excited-state hyperfine transitions to be resolved. As density increases the lines broaden. The excited-state hyperfine lines merge first, then as density further increases, the ground state hyperfine splitting can no longer be resolved. Zero on the detuning axis represents the weighted line centre of the D2 line

To highlight the line-broadening effect of dipole–dipole interactions, Fig. 5.6 shows the spectra on the Rb D2 line obtained in a cell with thickness $\lambda/2$ as the temperature (and hence atomic density) is tuned.

As temperature is increased, the evolution of the medium from a regime where dipole–dipole interactions are negligible ($T = 120$ °C, $N = 2 \times 10^{13}$ cm^{-3}, topmost line) to a regime where the interactions completely dominate the lineshape ($T = 330$ °C, $N = 5 \times 10^{16}$ cm^{-3}, bottom-most line) is clear. At the lowest densities, Dicke narrowing allows the excited state hyperfine levels to be resolved. As the density increases, the collisional linewidth increases linearly with density, and the hyperfine resolution is gradually lost. At a density of around $N \sim 10^{15}$ cm^{-3}, the collisional broadening becomes comparable to the excited state hyperfine splitting, and at $N \sim 10^{16}$ cm^{-3} the broadening is comparable to the ground state hyperfine splitting. After this the whole transition behaves essentially as a single S to P$_{3/2}$ transition (fine structure is still resolved, this splitting is around 7 THz).

5.4.1 Fitting Data

We fit the calibrated spectra to the model outlined in Chap. 3. The full susceptibility model is a function of four free parameters, and is characterised as

$$\chi(N, A, \Gamma_{\text{ad}}, \Delta_{\text{tot}}) = \frac{\mu^2}{\hbar\epsilon_0} \sum_{F, F'} C^2_{F, F'} \int \frac{iN_{\text{vc}}(v)}{\gamma - i\Delta'} \, dv, \tag{5.44}$$

with

$$\gamma = (\Gamma_0 + \beta N + \Gamma_{\text{ad}})/2,$$
$$\Delta' = \Delta + \Delta_{F, F'} + \Delta_{\text{vdW}} + \Delta_{\text{tot}} - kv,$$

and $N_{\text{vc}}(v)$ is given by Eq. (3.2). The $C^2_{F, F'}$ are the transition strengths of each individual hyperfine transition which have resonances at $\Delta_{F, F'}$. Δ_{vdW} accounts for the atom-surface induced shift (see Chap. 4). The susceptibility is related to the measured transmission lineshape, taking into account reflectivity effects, by Eq. (3.7).

The free parameters correspond to the atomic number density N; the amount of Dicke narrowing, characterised by the parameter A in the velocity distribution; an additional amount of (Lorentzian) line broadening, Γ_{ad} due to for example, atom-surface interactions; and finally a single shift parameter Δ_{tot}. The fit is implemented using a least-squares method with the Levenberg–Marquardt algorithm. The errors in the fit parameters are obtained from the covariance matrix, as outlined in Ref. [53].

Figure 5.7 shows two example data sets, at low and high density. The low density data (and green theory line) is for a cell thickness $\ell = \lambda/2$, at a temperature of $T = 130\,^\circ\text{C}$, which corresponds to a collectivity parameter $\mathcal{C} \approx 1$, and highlights the effects of Dicke narrowing at their most striking. In stark contrast to this, panel (b) shows a high density spectrum for $\ell = 90$ nm at a temperature $T = 305\,^\circ\text{C}$ ($\mathcal{C} \approx 300$), where dipole–dipole interactions dominate. For both data sets, we have good agreement between experiment and theory, as highlighted by the residuals, which are on the 1 % level. For the high density data, to highlight the line shift we have also plotted the theory curve with $\Delta_{\text{tot}} = 0$.

5.4.2 Saturation of Susceptibility

A consequence of the dipole–dipole interaction is that the susceptibility, and hence the optical depth, is no longer linearly dependent on the atomic density. and therefore the susceptibility has a density term in the denominator as well as the numerator

$$\text{Im}\,\chi(\Delta) = \frac{N\mu^2}{\hbar\epsilon_0} \frac{1}{(\Gamma_0 + \beta N)/2 - i(\Delta + \Delta_{\text{tot}})}. \tag{5.45}$$

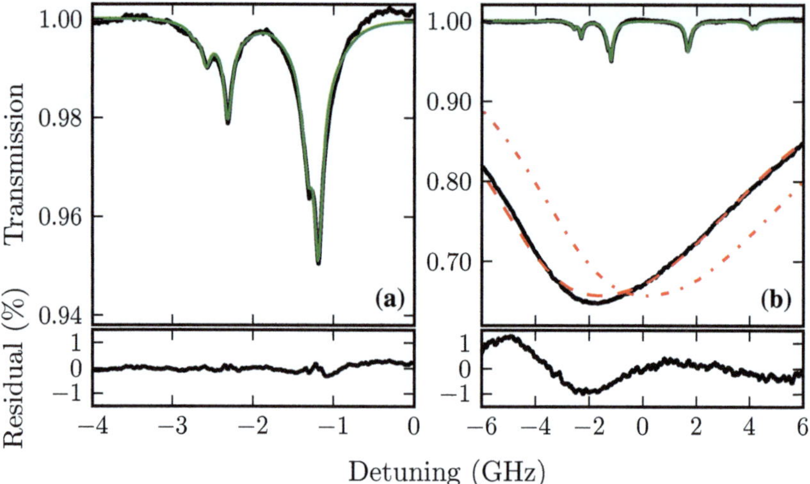

Fig. 5.7 Transmission spectra—experiment and theory. Transmission spectrum as a function of linear detuning for thickness. **a** $\ell = 390$ nm, $T = 130\,°\mathrm{C}$ ($\mathcal{C} \approx 1$), and **b** $\ell = 90$ nm, $T = 305\,°\mathrm{C}$ ($\mathcal{C} \approx 300$). The *black line* is experimental data, while the *solid green* and *dashed red lines* are the fits to the model. The *dotted red line* in panel (**b**) is the theory without the line shift included, to highlight its effect. The residuals show the difference between experiment and theory. Zero on the detuning axis represents the weighted line centre of the D2 line

On resonance, the susceptibility is purely imaginary, and at high density $\beta N \gg \Gamma_0$ so we can neglect the contribution from the natural linewidth. The density terms then cancel, yielding

$$\mathrm{Im}\ \chi(\Delta = 0) = \frac{2\mu^2}{\hbar\epsilon_0\beta}. \tag{5.46}$$

In addition, in the high density limit the linewidth becomes significantly larger than the splitting between the lines, so we can consider the whole D1(2) line a single transition from S to $\mathrm{P}_{1/2(3/2)}$. Substituting in for the self-broadening coefficient β [Eq. (5.37)], we find with linearly polarised light, the imaginary part of the resonant susceptibility saturates at a value

$$\mathrm{Im}\ \chi_{\mathrm{Sat}} = \sqrt{\frac{2J_e + 1}{2J_g + 1}} = \sqrt{2}, \tag{5.47}$$

for the D2 line, while for the D1 line the saturation value is 1. A similar saturation is found for the real part of the susceptibility, meaning that the refractive index also reaches a peak value, which will be the topic of the next chapter.

The saturation has profound implications for spectroscopic studies, as the opacity eventually reaches a peak value that only depends on the thickness of the medium. Physically, this can be seen as a consequence of the dipole–dipole interactions; as

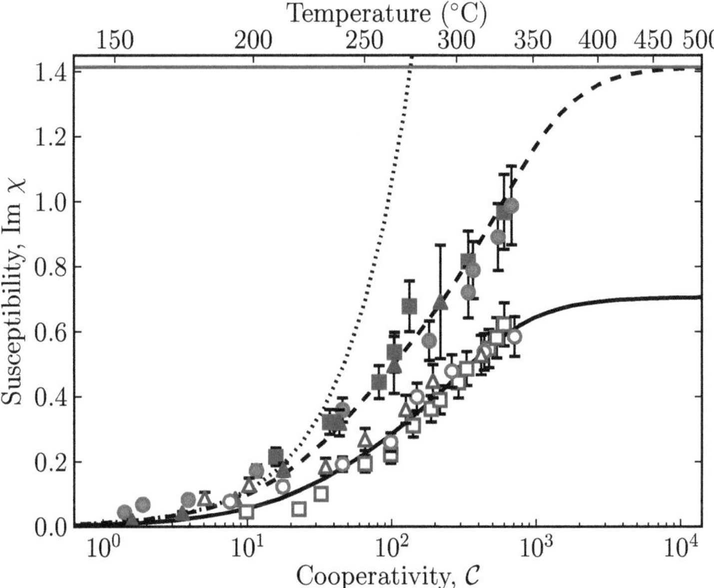

Fig. 5.8 Saturation of opacity due to dipole–dipole interactions. Without dipole–dipole interactions (*dashed line*) the opacity increases linearly with density. When interactions are included into the standard theory (*upper solid line*) the scaled opacity for the D_2 line is predicted to saturate at a value $N\sigma\lambdabar = \sqrt{2}$. However for thicknesses $\ell < \lambda/4$, we observe a change in the behaviour which doubles the broadening coefficient (*lower solid line*). A constant additional broadening of $\Gamma_{ad} = 2\pi \times 200 MHz$ has been added to the line for $\ell < \lambda/4$ account for atom–surface effects, which does not affect the value at which the opacity saturates at high density. Experimental data: $\ell = 90$ nm (*open red squares*); $\ell = 140$ nm (*open green circles*); $\ell = 180$ nm (*open blue triangles*); $\ell = 220$ nm (*closed red squares*); $\ell = 390$ nm (*closed green circles*); $\ell = 780$ nm (*closed blue triangles*). Error bars assume a 2 % error in transmission

more atoms are added into a volume, the extra absorption is exactly cancelled by an increase in the linewidth, which acts to reduce the effective resonant cross-section of each atom. Qualitatively this can be seen in Fig. 5.6, as the medium never becomes optically thick. Within the binary approximation, for $\ell = 390$ nm the transmitted intensity reaches a minimum $\mathcal{T}_{\min} = e^{-\sqrt{2}k\ell} = e^{-\sqrt{2}\pi} = 1.5\%$. To investigate the saturation quantitatively, we look at how the opacity scales with number density. We can extract the susceptibility from the transmission data in accordance with Beer's law. In Fig. 5.8 we plot the susceptibility at a detuning $\Delta \approx -1.2$ GHz (the unshifted resonance position of the ^{85}Rb $F_g = 3 \rightarrow F_e = 4$ transition, though in principle one can choose any detuning and achieve similar results) as a function of the collectivity parameter, for different cell thicknesses. The advantage of this approach is that there is no fitting involved.

If there was no density dependent line broadening, the data would be expected to follow the dotted line, but there is clear deviation from this. For thicknesses $\ell > \lambda/4$,

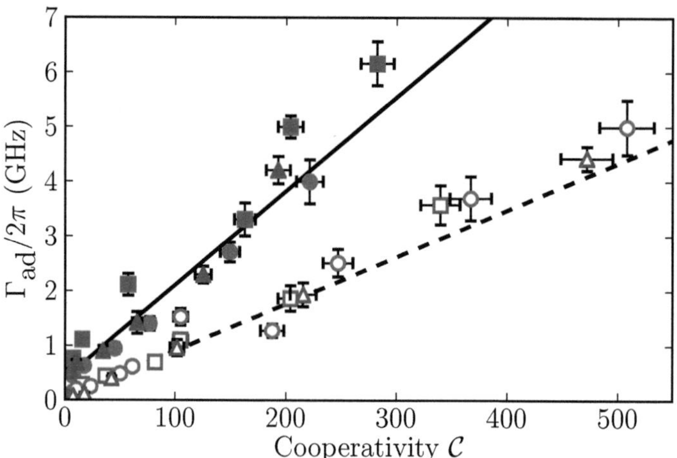

Fig. 5.9 Fitted broadening coefficient Γ_{ad} as a function of the collectivity parameter, for the same vapour thicknesses (and corresponding symbols) as Fig. 5.8. The *dashed line* has a gradient of β and zero offset, whilst the *solid line* has slope 2β and an offset of 200 MHz

the data lie on the dashed curve that follows conventional theory, saturating at $\sqrt{2}$. For thicknesses $\ell < \lambda/4$ however, the data do not follow this curve. They still follow a saturation curve, but calculated with twice the amount of self-broadening, and hence the saturating value is lower by a factor of 2.

As an additional method of analysis, we also fit the data to the susceptibility model, and extract a broadening coefficient, excluding the βN term in the linewidth, so that self-broadening is now included into the fit parameter Γ_{ad}.

In Fig. 5.9 we plot the fitted value of Γ_{ad} as a function of density, for the same cell thicknesses as plotted in Fig. 5.8. Again, the same trend is visible; for $\ell > \lambda/4$, the data agrees with conventional self-broadening theory, following a straight line with gradient β. However, for $\ell < \lambda/4$, the data fit to a straight line but with a significantly larger gradient. In the same way as for Fig. 5.8, the solid line in the figure represents a gradient of 2β, with an additional offset to account for atom-surface broadening.

We can further analyse the data in Figs. 5.8 and 5.9 and extract a line broadening parameter by fitting to the curves allowing the interaction strength β to vary. We extract a broadening coefficient representing the strength of the dipole–dipole interaction as a function of cell thickness, plotted in Fig. 5.10. The offset, which we interpret as the atom-surface induced broadening, is presented in Fig. 4.6 in the previous chapter. We find a sharp transition between conventional self-broadened lines and the enhanced broadening at $\ell = \lambda/4 = 195$ nm. Below $\lambda/4$, the data fit with a broadening coefficient $\Gamma/N_0 = (2.07 \pm 0.06)\beta$, while for larger thicknesses we find $\Gamma/N_0 = (0.94 \pm 0.05)\beta$. The exact mechanism for this apparent enhancement is not well understood and requires further investigation. Observation of off-axis fluorescence has shown promise as an experimental technique, yielding higher resolution

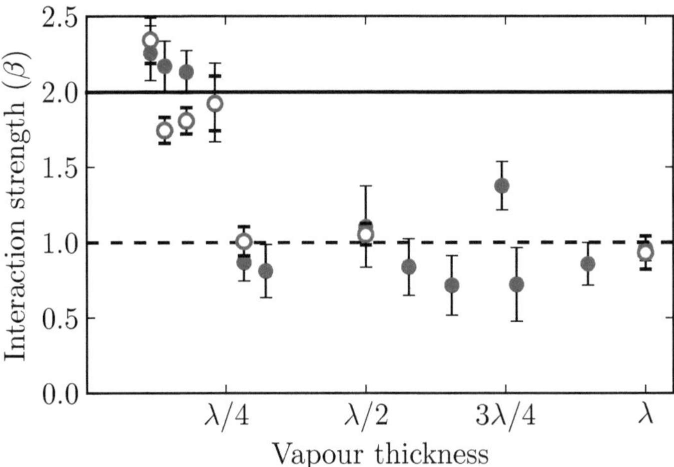

Fig. 5.10 Dipole–dipole interaction strength, as measured by the broadening coefficient, in units of β, plotted as a function of layer thickness. Closed symbols are calculated from fitting the data shown in Fig. 5.8 allowing the broadening coefficient to vary, whilst open symbols are calculated from fitting the data in Fig. 5.9. We calculate the weighted average interaction strength for $\ell < \lambda/4$ to be $(2.07 \pm 0.06)\beta$, whereas for $\ell > \lambda/4$ we find $(0.94 \pm 0.05)\beta$

of the line widths at moderate densities—an example spectrum using this method is shown in Appendix C.

5.4.3 Cooperative Lamb Shift

Whilst Fig. 5.6 clearly shows the evolution from independent to collective response, the line profile becomes very flat at high density (due to the large optical depth) and therefore the shift of the lines is less apparent. To highlight the line shift in more detail, in Fig. 5.11 we show a similar spectral evolution, for a thickness $\ell = 90$ nm where as we will show later the density-dependent shift is much larger. Unlike the data at $\ell = \lambda/2$, even for the lowest densities there is no resolution of the excited state hyperfine levels, since the broadening due to the atom-surface interaction is already larger than the hyperfine splitting.

The shift is immediately apparent from the figure, and although by eye one might imagine that the line shift follows the point of minimum transmission, this is not the case. As has already been shown in Fig. 5.7, if the shift is not included in the model the peak absorption at high density occurs at a detuning $\Delta \approx 0$. Given this, it becomes clear that the highest density spectra is shifted by around 2 GHz.

Again we fit the data to the model and extract the shift parameter. The extracted shift as a function of number density is shown in Fig. 5.12 for three thicknesses, $\ell = 90, 250$ nm and 420 nm. In the high density regime ($N > 10^{16}$ cm^{-3}) dipole–dipole

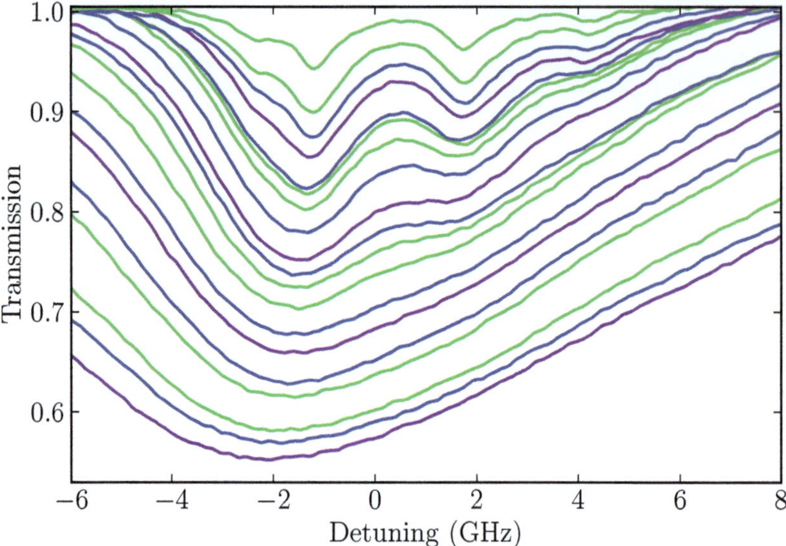

Fig. 5.11 Evolution of the transmission spectrum for $\ell = 90$ nm showing significant red-shift of the absorption lines at high density. The lowest and highest temperatures are $T = 180\,^{\circ}$C and $T = 315\,^{\circ}$C. As in Fig. 5.6, as density increases there is significant broadening of the spectral lines. At low density however, atom-surface interactions already contribute significantly to spectral broadening so the excited-state hyperfine resolution from Dicke narrowing is washed out. At high density the shift is $\sim$2 GHz, as the centre-of-mass of the absorption line lies at zero detuning

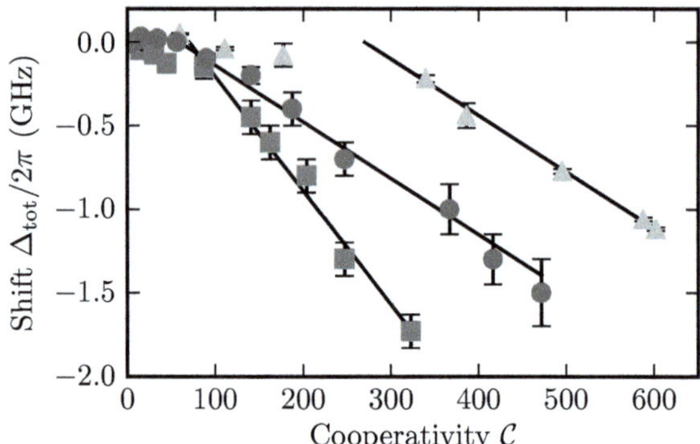

Fig. 5.12 Shift of resonance lines with density. Measured shift of resonance lines with density and fit to the linear, high density region for $\ell = 90$ nm (*red squares*) and $\ell = 250$ nm (*blue circles*). All other data sets follow the same trend - the linear region starts at $\mathcal{C} \approx 100$, except for the anomalous data set at $\ell = 420$ nm (*green triangles*), which clearly does not follow this trend. Possible reasons for the deviation are discussed in the text

interactions dominate the lineshape and we can treat the line as a single $S_{1/2} \to P_{3/2}$ transition which shifts linearly with density. For nearly all thicknesses the shift starts to become linear at $C \sim 100$ ($N \sim 0.75 \times 10^{16}$ cm^{-3}). However, for a thickness around $\ell = 420$ nm we observe a deviation from this, where the linear region does not appear to begin until much higher in density. In addition, the gradient is not what is expected from the CLS curve. We attribute this anomalous behaviour to the 5S-6P atomic resonance which has a wavelength of 420 nm, and can be populated by the well-known energy pooling process [54]. For this reason this data point has been omitted from Fig. 5.13.

We fit the gradient of the linear region to obtain the coefficient of the shift Δ_{tot}/N, and repeat these measurements for several thicknesses up to 600 nm. For thicknesses greater than 600 nm, the high optical depth of the sample impairs resolution of the line shift.

In Fig. 5.13 we plot Δ_{tot}/N as a function of cell thickness.

We extract the collisional shift by comparing the data to Eq. (5.43) with Δ_{col} the only free parameter. The amplitude and period of the oscillatory part are fully constrained by Eq. (5.42). We find the collisional shift to be $\Delta_{\text{col}}/2\pi = -(0.25 \pm 0.01) \times 10^{-7}$ Hz cm^3, similar to previous measurements on potassium vapour [30].

Though there is expected to be some density dependence of the atom-surface interaction, extrapolating from data in Ref. [55], the gradient is of the order of 10^{-9} Hz cm^3, nearly two orders of magnitude smaller than the measured gradients in Fig. 5.13, and so the atom-surface interaction can be neglected.

The solid line is the prediction of Eq. (5.42), in SI units on the left axis and scaled units on the right axis. The agreement between the measured shifts and the theoretical prediction is remarkable (the reduced χ^2 for the data is 1.7, indicating a good fit [53]). As well as measuring the thickness dependence of the CLS, our data also provide a determination of the Lorentz shift which can only be measured in the limit of zero thickness.

The coloured areas on the plot highlight the various contributions to the total shift. The red line is the constant Lorentz shift, the yellow shaded area is the collisional shift, which is density dependent but independent of the thickness of the medium. Finally the blue area, highlights that the CLS is dependent on both the density and thickness of the medium.

The observation of the CLS in a high density thermal vapour is an exciting result not only in itself, but because of its implications for observing other coherent interactions in the medium. The CLS is derived in the limit of a static medium, and in a thermal vapour motional effects are usually the dominant dephasing mechanism. We would therefore expect any coherent effect like the CLS to be completely washed out by the atomic motion.

The fact that we still observe this coherent response is encouraging (and somewhat surprising!) as it means that motional dephasing is not the dominant mechanism. We find experimentally that the line starts to shift linearly once the density reaches around 10^{16} cm^{-3}. If we use the inverse of the interaction-induced linewidth $\tau \sim (\beta N)^{-1}$ as a guide to the timescales, at this density $\tau \sim 1$ ns. Given a mean atomic speed of 300 nm/ns (which varies only very weakly with temperature), the atoms still move

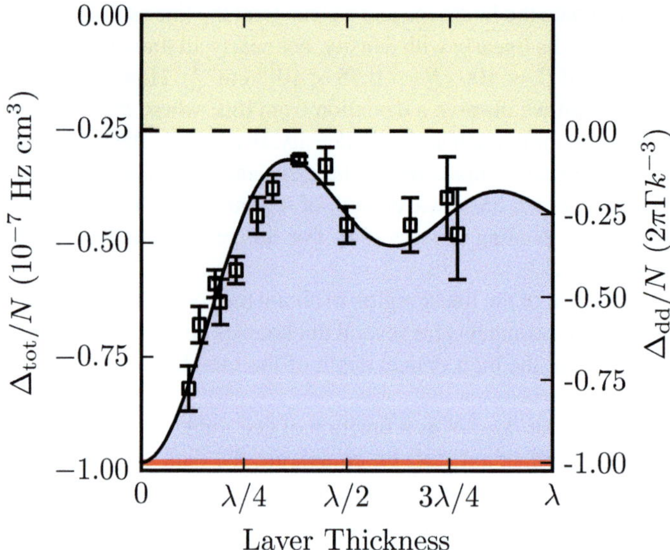

Fig. 5.13 Experimental verification of the cooperative Lamb shift. The gradient of the shift Γ_{tot}/N is plotted against cell thickness ℓ. The *solid black line* is Eq. (5.43) with $\Delta_{\mathrm{col}}/2\pi = -0.25 \times 10^{-7}$ Hz cm^3 and no other free parameters. The coloured areas highlight the different contributions to Δ_{tot}; the Lorentz shift (*red line*), the CLS (*blue*), and the collisional shift (*yellow*). The alternate ordinate axis shows Δ_{dd} in scaled units

a significant distance compared to the wavelength, so it is surprising that motional effects are not more important here. As τ becomes smaller at even larger densities the picture of a frozen medium becomes more plausible, as the atoms no longer move significantly during the interaction. At the highest densities in this work the linewidth is of the order of 10 GHz, which implies that the corresponding interaction time is very short, of the order of 0.1 ns. A typical atom only moves around 30 nm in this time, much smaller than the wavelength, so a quasistatic picture is likely to be more valid here.

5.5 Outlook

The demonstration of the CLS and coherent dipole–dipole interactions in sub-wavelength thickness thermal vapours is an important result, opening a new domain for the investigation of coherent interactions in experimentally simple systems.

We have shown that with strong enough interactions, which can be easily realised by tuning the temperature of the vapour, motional effects that are usually the dominant dephasing process in thermal vapours can be overcome. This allows the observation of the coherent character of the dipole–dipole interaction, which in our system manifests as a shift of the spectral lines.

We also observe an apparent enhancement in the decay rate which we measure as an increase in the line broadening as the layer thickness is reduced below $\lambda/4$. The mechanism for this process is not understood and requires further investigation.

As the CLS depends on the degree of excitation [21], exotic non-linear effects such as mirrorless bistability [4, 56] may be accessible experimentally. Such effects have recently been observed in thermal vapours with Rydberg atoms [57], and it remains an interesting question to whether, and to what extent, these effects are observable with ground-state atomic vapours.

It is also important to note that the observed CLS is still an angular average, that could be optimised and controlled by changing the geometry of the cell. Whilst the sub-wavelength thickness of the vapour provides one degree of confinement, an enhanced effect could be engineered by confining the dipoles even further. Three-dimensional confinement using nanostructures has recently been demonstrated with Cs atoms [58]. Alternately, one could imagine that by lithographically etching out channels in one of the cell windows into thin rows, for example, that the atoms could be confined into pseudo-1D strips. In this case the angular distribution of the dipoles could be tightly confined and an enhanced or reduced interaction could be possible, by changing the polarisation of the incoming light.

References

1. R.G. DeVoe, R.G. Brewer, Observation of superradiant and subradiant spontaneous emission of two trapped Ions. Phys. Rev. Lett. **76**, 2049 (1996)
2. C. Hettich et al., Nanometer resolution and coherent optical dipole coupling of two individual molecules. Science **298**, 385 (2002)
3. J. Javanainen, J. Ruostekoski, Y. Li, S.-M. Yoo, Shifts of a resonance line in a dense atomic sample. Phys. Rev. Lett. **112**, 113603 (2014)
4. M. Fleischhauer, S. Yelin, Radiative atom–atom interactions in optically dense media: quantum corrections to the Lorentz–Lorenz formula. Phys. Rev. A **59**, 2427 (1999)
5. A.P. Thorne, *Spectrophysics*, 2nd edn. (Chapman and Hall, London, 1988)
6. E. Lewis, Collisional relaxation of atomic excited states, line broadening and interatomic interactions. Phys. Rep. **58**, 1 (1980)
7. C.A. Brau, *Modern Problems in Classical Electrodynamics* (OUP, Oxford, 2004)
8. P. Siddons, C.S. Adams, C. Ge, I.G. Hughes, Absolute absorption on rubidium D lines: comparison between theory and experiment. J. Phys. B **41**, 155004 (2008)
9. M.J. Stephen, First-order dispersion forces. J. Chem. Phys. **40**, 669 (1964)
10. J.D. Jackson, *Classical Electrodynamics*, 3rd edn. (Wiley, Hoboken, 1999)
11. D.-W. Wang, Z.-H. Li, H. Zheng, S.-Y. Zhu, Time evolution, Lamb shift, and emission spectra of spontaneous emission of two identical atoms. Phys. Rev. A **81**, 043819 (2010)
12. R. Dicke, Coherence in spontaneous radiation processes. Phys. Rev. **93**, 99 (1954)
13. B.H. Bransden, C.J. Joachain, *Physics of atoms and molecules*, 2nd edn. (Pearson Education, Harlow, 2003)
14. L. Weller, R.J. Bettles, P. Siddons, C.S. Adams, I.G. Hughes, Absolute absorption on the rubidium D1 line including resonant dipoledipole interactions. J. Phys. B **44**, 195006 (2011)

15. S.L. Kemp, I.G. Hughes, S.L. Cornish, An analytical model of off-resonant Faraday rotation in hot alkali metal vapours. J. Phys. B **44**, 235004 (2011)
16. L. Weller, T. Dalton, P. Siddons, C.S. Adams, I.G. Hughes, Measuring the Stokes parameters for light transmitted by a high-density rubidium vapour in large magnetic fields. J. Phys. B **45**, 055001 (2012)
17. L. Weller et al., Absolute absorption and dispersion of a rubidium vapour in the hyperfine Paschen–Back regime. J. Phys. B **45**, 215005 (2012)
18. L. Weller et al., Optical isolator using an atomic vapor in the hyperfine Paschen–Back regime. Opt. Lett. **37**, 3405 (2012)
19. A. Reinhard, T. Liebisch, B. Knuffman, G. Raithel, Level shifts of rubidium Rydberg states due to binary interactions. Phys. Rev. A **75**, 032712 (2007)
20. R.W. Boyd, *Nonlinear Optics*, 3rd edn. (Academic Press, USA, 2008)
21. R. Friedberg, S.R. Hartmann, J.T. Manassah, Frequency shifts in emission and absorption by resonant systems ot two-level atoms. Phys. Rep. **7**, 101 (1973)
22. H.A. Lorentz, *The Theory of Electrons* (BG Teubner, Leipzig, 1909)
23. M. Born, E. Wolf, *Principles of Optics*, 7th edn. (CUP, Cambridge, 1999)
24. S. Chandrasekhar, Stochastic problems in physics and astronomy. Rev. Mod. Phys. **15**, 1 (1943)
25. H. Foley, The pressure broadening of spectral lines. Phys. Rev. **69**, 616 (1946)
26. A. Corney, *Atomic and Laser Spectroscopy* (OUP, Oxford, 1989)
27. J. Woerdman, F. Blok, M. Kristensen, C. Schrama, Multiperturber effects in the Faraday spectrum of Rb atoms immersed in a high-density Xe gas. Phys. Rev. A **53**, 1183 (1996)
28. V.A. Sautenkov, H. van Kampen, E. Eliel, J. Woerdman, Dipole–dipole broadened line shape in a partially excited dense atomic gas. Phys. Rev. Lett. **77**, 3327 (1996)
29. H. van Kampen et al., Observation of collisional modification of the Zeeman effect in a high-density atomic vapor. Phys. Rev. A **56**, 310 (1997)
30. J.J. Maki, M.S. Malcuit, J.E. Sipe, R.W. Boyd, Linear and nonlinear optical measurements of the Lorentz local field. Phys. Rev. Lett. **67**, 972 (1991)
31. M.O. Scully, A.A. Svidzinsky, The effects of the N atom collective Lamb shift on single photon superradiance. Phys. Lett. A **373**, 1283 (2009)
32. R. Röhlsberger, K. Schlage, B. Sahoo, S. Couet, R. Rüffer, Collective Lamb shift in single-photon superradiance. Science **328**, 1248 (2010)
33. W. Lamb, R. Retherford, Fine structure of the hydrogen atom by a microwave method. Phys. Rev. **72**, 241 (1947)
34. R. Friedberg, J.T. Manassah, Analytic expressions for the initial cooperative decay rate and cooperative Lamb shift for a spherical sample of two-level atoms. Phys. Lett. A **374**, 1648 (2010)
35. R. Friedberg, J.T. Manassah, Initial cooperative decay rate and cooperative Lamb shift of resonant atoms in an infinite cylindrical geometry. Phys. Rev. A **84**, 023839 (2011)
36. M.O. Scully, Collective Lamb shift in single photon dicke superradiance. Phys. Rev. Lett. **102**, 143601 (2009)
37. T. Bienaimé, M. Petruzzo, D. Bigerni, N. Piovella, R. Kaiser, Atom and photon measurement in cooperative scattering by cold atoms. J. Mod. Opt. **58**, 1942 (2011)
38. T. Bienaimé, R. Bachelard, N. Piovella, R. Kaiser, Cooperativity in light scattering by cold atoms. Fortschr. Phys. **61**, 377 (2013)
39. W.R. Garrett, R.C. Hart, J.E. Wray, I. Datskou, M.G. Payne, Large multiple collective line shifts observed in three-photon excitations of Xe. Phys. Rev. Lett. **64**, 1717 (1990)
40. R. Röhlsberger, The collective Lamb shift in nuclear γ-ray superradiance. J. Mod. Opt. **57**, 1979 (2010)
41. M. Gross, S. Haroche, Superradiance: an essay on the theory of collective spontaneous emission. Phys. Rep. **93**, 301 (1982)
42. N. Skribanowitz, I. Herman, J. MacGillivray, M. Feld, Observation of Dicke superradiance in optically pumped HF gas. Phys. Rev. Lett. **30**, 309 (1973)
43. M. Gross, C. Fabre, P. Pillet, S. Haroche, Observation of near-infrared Dicke superradiance on cascading transitions in atomic sodium. Phys. Rev. Lett. **36**, 1035 (1976)

44. S. Inouye, Superradiant rayleigh scattering from a Bose–Einstein condensate. Science **285**, 571 (1999)
45. T. Bienaimé, N. Piovella, R. Kaiser, Controlled Dicke subradiance from a large cloud of two-level systems. Phys. Rev. Lett. **108**, 123602 (2012)
46. M.O. Scully, E.S. Fry, C.H.R. Ooi, K. Wódkiewicz, Directed spontaneous emission from an extended ensemble of N atoms: timing is everything. Phys. Rev. Lett. **96**, 010501 (2006)
47. J.T. Manassah, Giant cooperative Lamb shift in a density-modulated slab of two-level atoms. Phys. Rev. A **374**, 1985 (2010)
48. H. Li, V.A. Sautenkov, Y.V. Rostovtsev, M.O. Scully, Excitation dependence of resonance line self-broadening at different atomic densities. J. Phys. B **42**, 065203 (2009)
49. V.A. Sautenkov, Line shapes of atomic transitions in excited dense gas. Laser Phys. Lett. **8**, 771 (2011)
50. M.D. Lukin et al., Dipole blockade and quantum information processing in mesoscopic atomic ensembles. Phys. Rev. Lett. **87**, 037901 (2001)
51. M. Saffman, T. Walker, K. Mølmer, Quantum information with Rydberg atoms. Rev. Mod. Phys. **82**, 2313 (2010)
52. R. Friedberg, S. Hartmann, J. Manassah, Mirrorless optical bistability condition. Phys. Rev. A **39**, 3444 (1989)
53. I.G. Hughes, T.P.A. Hase, *Measurements and their Uncertainties: A Practical Guide to Modern Error Analysis* (OUP, Oxford, 2010)
54. L. Barbier, M. Cheret, Energy pooling process in rubidium vapour. J. Phys. B **16**, 3213 (1983)
55. I. Hamdi et al., Laser spectroscopy with nanometric gas cells: distance dependence of atom surface interaction and collisions under confinement. Laser Phys. **15**, 987 (2005)
56. C.M. Bowden, C.C. Sung, First- and second-order phase transitions in the Dicke model: relation to optical bistability. Phys. Rev. A **19**, 2392 (1979)
57. C. Carr, R. Ritter, K.J. Weatherill, C.S. Adams, Cooperative non-equilibrium phase transition in a dilute thermal atomic gas (2013), arxiv.org, 1302.6621, arXiv:1302.6621v1
58. P. Ballin, E. Moufarej, I. Maurin, A. Laliotis, D. Bloch, Three-dimensional confinement of vapor in nanostructures for sub-Doppler optical resolution. Appl. Phys. Lett. **102**, 231115 (2013)

Chapter 6
Giant Refractive Index

In the previous chapter, the effects of dipole–dipole interactions were explored, leading to the saturating behaviour of the electric susceptibility at high density. Experimentally this was demonstrated via a saturation in the optical depth of the ensemble. However, this only probes the imaginary component of the susceptibility. In this chapter we investigate the real part of the susceptibility, responsible for the refractive properties of the medium.

6.1 Introduction

We should, at this point, clarify what is meant by the word 'giant' in the title of this chapter. In solid-state physics, an entirely new field has recently developed based on so-called 'metamaterials' [1]; materials made up from nano-scale building blocks, engineered to have drastically different properties to most other materials found in nature. A subset of these are materials with negative refractive indices [2], which change the direction of light in the opposite direction to every other material found in nature. By engineering the relative permittivity and permeability of these metamaterials it has been possible to heavily modify the refractive index, recently achieving a peak refractive index $n = 38.6$ in the THz frequency domain [3]. Construction of metamaterials for optical wavelengths is also possible; see for example [4]. There has been a great deal of recent work using these metamaterials to engineer 'invisibility cloaks', in the microwave [5] and optical frequency domains [6, 7].

In this chapter we show that the refractive index of a Rb vapour can be as high as $n = 1.3$. While this is clearly significantly lower than what can be realised in the solid-state, this is realised with a vapour with a pressure two orders of magnitude lower than atmospheric pressure.[1] In terms of atomic vapour experiments, our index enhancement $n - 1$ is three orders of magnitude larger than has been observed previously for a ground state transition [9].

---

[1] For air, the index change at optical frequencies $n - 1 \approx 3 \times 10^{-4}$ [8].

J. Keaveney, *Collective Atom–Light Interactions in Dense Atomic Vapours*,
Springer Theses, DOI: 10.1007/978-3-319-07100-8_6,
© Springer International Publishing Switzerland 2014

6.1.1 Phase Shift and Refractive Index

As mentioned in the previous chapter, the fundamental process of light scattering is the interference between incident and dipolar fields. The dipolar field is that of a driven oscillator, and so has a different (frequency-dependent) phase to the input field. The output is just the superposition and is therefore also phase-shifted. Let's neglect the spatial parts of the field, leaving only the oscillatory parts, and furthermore assume that the laser is near-resonance, so the laser frequency $\omega_p \approx \omega_0$, the atomic resonance frequency. The transmitted field is then

$$E_{\mathrm{out}} = |E_{\mathrm{in}}| \cos(\omega_0 t) + |E_{\mathrm{dipole}}| \cos(\omega_0 t + \phi(\Delta))$$
$$\propto \cos(\omega_0 t + \Phi). \tag{6.1}$$

If the magnitudes of the two fields are equal then this has the analytic solution $E_{\mathrm{out}} = 2|E| \sin(\omega_0 t + \phi/2) \sin(\phi/2)$; in general Φ is dependent on $|E_{\mathrm{in}}|$, $|E_{\mathrm{dipole}}|$ and ϕ, which depends on the detuning from resonance Δ. In a macroscopic medium, we can build up a total phase difference by combining many layers each with a phase offset Φ. If phase lags behind the driving field, $\Phi < 0$, points of constant phase travel through the medium slower than through vacuum, and hence we have a phase velocity $v = c/n$ with $n > 1$.

Over a medium with thickness ℓ, the phase change that is built up is related to the real part of the electric susceptibility

$$\Phi = \mathrm{Re}\left(\sqrt{1+\chi}\right) k\ell, \tag{6.2}$$

where as usual $k = 2\pi/\lambda$ is the wavevector of the light field. We have already seen in the previous chapter that the imaginary part of the susceptibility is greatly modified by dipole–dipole interactions between atoms, leading to remarkable effects in the absorption of the incident light field. In this chapter we investigate how the same physics affects the real part of the susceptibility and hence the refractive index of the medium.

6.2 Experimental Setup

Figure 6.1 shows the experimental setup. We employ a heterodyne interferometer similar to that used in [10]. An acousto-optic modulator (AOM) is double-passed to generate a second beam at $\Delta_{\mathrm{AOM}}/(2\pi) = 160\,\mathrm{MHz}$, which is power-matched and then recombined with the unshifted beam. After recombination at the polarising beam splitter (PBS) cube the beams are of different linear polarisations; a half-waveplate set at $45°$ followed by another PBS cube matches them, at the cost of half of the light which is rejected at the beam splitter. The beams are coupled into a single-mode optical fibre and are therefore mode-matched on exiting the fibre. Half

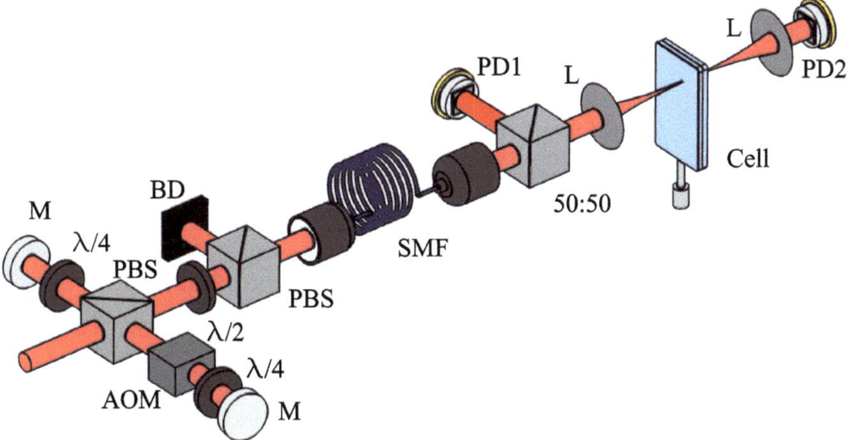

Fig. 6.1 Experimental setup figure and raw data example. *L*—lens, *M*—mirror, $\lambda/4, \lambda/2$—quarter/half waveplates, *AOM*—acousto-optic modulator, *SMF*—single mode fiber, *PBS*—polarising beamsplitter, *50:50*—non-polarising beamsplitter, *PD1/2*—fast photodiodes, *BD*—beam dump

of the light is then sent to a fast photodiode (PD1) and the other half focussed to a spot size ($1/e^2$ radius) of 50 μm through the cell and onto a second fast photodiode (PD2). A relatively large beam waist was chosen as the fast photodiodes (model EOT-2030A) require more incident optical power than the ones used in the transmission spectroscopy experiments. In order to have a detectable signal and avoid optical pumping effects (see Sect. 3.3) the beam waist must be increased. Typically the power incident on the cell was $\approx$30 μW (intensity $I \approx 100\,\mathrm{mW/cm^2}$). The larger focal spot leads to a thickness variation of around 10 nm due to the geometry of the cell.

Both detectors measure the 160 MHz beat frequency of the two light fields. In a single acquisition we observe $\sim$300 oscillations over 2 μs. We take 100 measurements of the relative phase at each detuning point, from which a mean and standard error are calculated. An example of the two beat signals and the measured relative phase are shown in Fig. 6.2. The measurements are taken using the oscilloscope's built-in functions (for speed of data acquisition) and so there is no error bar on an individual measurement. The method has been checked against a separate analysis routine, fitting sine curves to the raw data and extracting a phase difference, as shown in Fig. 6.3. The two methods agree to within 1 %.

The laser is then tuned across the D2 resonance line in steps of 100 MHz. For each value of the laser detuning, the laser is stabilised ('locked') electronically to within 10 MHz (error bars in this axis are not shown in the figures) using a HighFinesse wavemeter and a proportional/integral/derivative (PID) feedback loop implemented in LabView. Whilst the lock is not as stable as using a spectroscopic locking method, the key advantage is that it is possible to lock to an arbitrary frequency, as opposed to being limited to locking around atomic resonances.

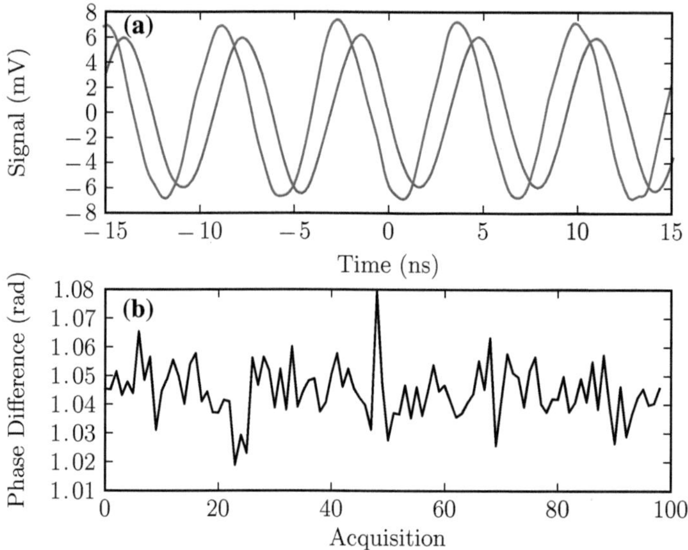

Fig. 6.2 **a** Example data of the beat frequencies of the shifted (*blue*) and unshifted (*red*) beams from which a frequency-dependent phase difference is measured. **b** This is repeated for 100 acquisitions, from which a mean and standard error are calculated

Since the two beams are at nearly the same frequency ($\sim$380 THz frequency with a 160 MHz difference) and travel down the same optical paths, the phase difference between the beat signals measured at the two detectors is given by

$$\phi = \phi_0 + \Delta\phi = \phi_0 + (n(\Delta + \Delta_{\text{AOM}}) - n(\Delta))\,k\ell, \tag{6.3}$$

where ϕ_0 is an unimportant global phase due to different optical path lengths to the two detectors. Essentially the measurement is a measurement of the refractive index gradient. We then reconstruct the refractive index profile from fitting the measured relative phase $\Delta\phi$ to the weak-probe model outlined previously. For relatively high atomic number densities where the homogeneous broadening is larger than Δ_{AOM}, we can simultaneously measure the transmitted intensity by measuring the amplitude of the oscillations.

As mentioned in Ref. [10], the heterodyne interferometry technique is very insensitive to vibrations, as the light beams see the same optical path. Though the global phase may vary due to vibrations, the only relative phase shift (dependent on laser frequency) comes from the atoms. This allows us to achieve sensitivities on the order of 1 mrad.

The interferometry technique also has advantages over conventional transmission spectroscopy, as much of the calibration (see Appendix B) is not needed, and there are no additional effects due to the cell reflections. The main disadvantage is for high optical depths, as the precision of the phase measurement is affected by the amount of light reaching the detector.

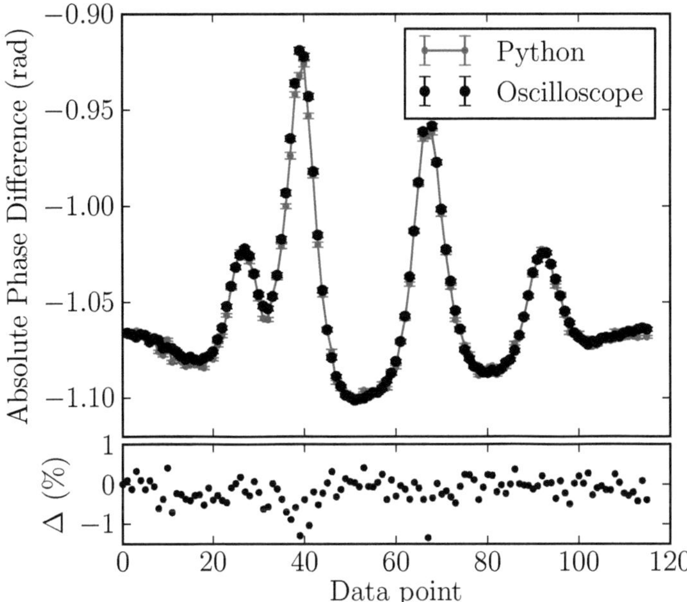

Fig. 6.3 Comparison of measured phase data, between (*black*) using the oscilloscope function, averaged over 100 acquisitions, and (*red*) post-processed data using a fitting routine written in Python. The *bottom* panel shows the difference between the two methods, which agree to within 1 %

6.3 Maximum Refractive Index

An interesting question that arises when considering an index enhancement this large is what is the maximum possible index of the medium?

Whilst previously it has been speculated that the near-resonance refractive index of a gaseous medium could be "as high as 10 or 100" [11], this estimate was based on an independent dipole model and neglected the dipole–dipole interactions which are invariably present at high density. As we have seen in Chap. 5, eventually the increase in the susceptibility due to adding more dipoles is cancelled out by the higher damping rate due to resonant dipole–dipole interactions [12, 13]. Since the refractive index of the medium is related to the real part of the susceptibility, the index also saturates. A maximum index of $n \approx 1.4$ has previously been predicted for Rb [14]. Using our susceptibility model, within the limits of the binary approximation we calculate the maximum refractive index to be $n = 1.31$. This occurs at $T \sim 360\,^{\circ}\mathrm{C}$ at a detuning $\Delta \sim -7\,\mathrm{GHz}$ from the weighted line center, as shown in Fig. 6.4. The exact detuning depends on how the lines shift, so will depend on the thickness of the vapour; here we have calculated in the limit of large vapour thickness, where the sinusoidal oscillations have damped out. For comparison, a 75 mm cell at room temperature (using the same theory as above) would have a maximum

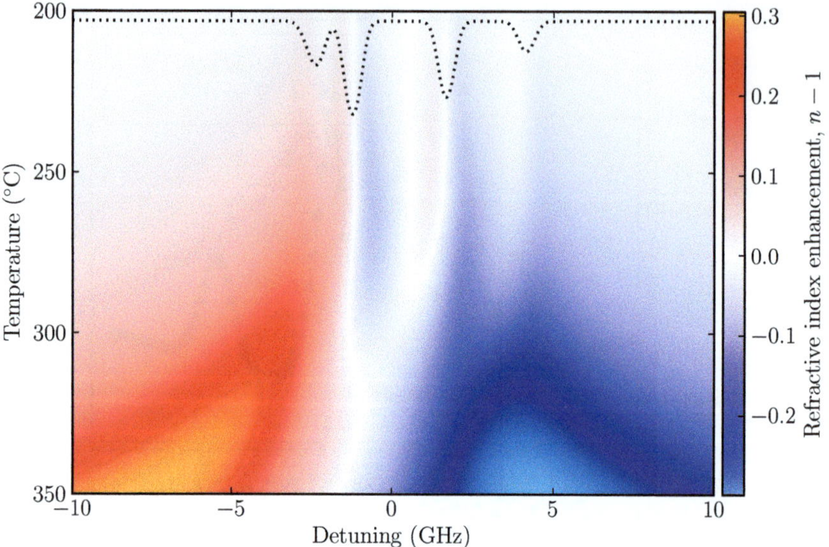

Fig. 6.4 Calculated refractive index as a function of temperature and detuning over the D2 resonance line. The maximum predicted index is 1.31. The density dependent redshift can clearly be seen, and has a significant effect on the position of the maximum index (we take $\ell \gg \lambda$ for this calculation)

index enhancement $n - 1 \approx 4 \times 10^{-7}$, but this still results in a significant phase shift, since $\Delta\phi = \Delta n k \ell$ and the cell thickness ℓ is orders of magnitude larger.

The intimate link between refraction and absorption manifest in the Kramers–Kronig relations make the observation of a large index enhancement challenging, since a large index is accompanied by high absorption. For the highest temperatures we can reach with the current experimental setup ($T = 360\,°\mathrm{C}$, $N \approx 10^{17}\,\mathrm{cm}^{-3}$), the vapour thickness must be much less than the wavelength of light in order to still transmit some light. However, as one reduces cell thickness atom-wall interactions start to have a significant effect on the linewidth, introducing further undesirable broadening and shifts which reduce the maximum observable index. In practice we are forced to compromise slightly; Fig. 6.5 shows experimental data with thickness $\ell = 250\,\mathrm{nm}$, at $T = 330\,°\mathrm{C}$. For these conditions, the high on resonance optical depth reduces signal to the point where a phase cannot be accurately measured, resulting in larger experimental uncertainty. Away from resonance, however, we observe good agreement with theory. From this we infer a maximum refractive index $n = 1.26 \pm 0.02$ approximately 5 GHz red detuned of line centre. The error bar is derived from the fit parameter Γ_{ad}.

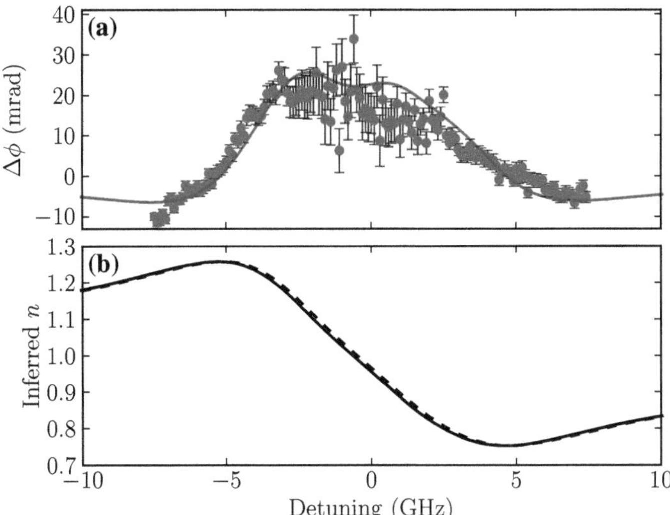

Fig. 6.5 Experimental data for a thickness $\ell = 250\,\text{nm}$ and $T = 330\,^\circ\text{C}$ ($N = 5 \times 10^{16}\,\text{cm}^{-3}$). **a** Measured phase shifts and fit to model. **b** Inferred refractive index profile. The *solid line* is the unshifted beam, the *dashed line* is the beam detuned by Δ_{AOM}. Close to resonance, the large optical depth reduces the signal to the point where accurate phase information is lost. Despite this, the fit to theory is reasonable and we infer a maximum index $n = 1.26 \pm 0.02$

6.4 Phase Shift Due to a Single Atomic Layer

Due to the scalability of both the atomic density and vapour thickness, an interesting regime is available whereby the atomic separation can be greater than the cell thickness. In this regime the light field will only interact with a vapour layer that is on average one atom thick. This poses a fundamental question: at what point does a collection of atoms start to act as a bulk medium, and when do they need to be considered as individual emitters? With tight enough focussing and low enough atomic density, could we enter a regime where we have on average zero atoms in the beam, and could we detect the transit of a single atom? On this scale, exceptionally good mode matching is required between the incident and atomic fields to achieve significant extinction of the incident field. In principle perfect cancellation of the incident field is possible; in this case the atom acts like a mirror, reflecting back all of the incident field and cancelling it completely in the forward direction [15]. Mode matching the atomic field (a spherical wave) with the focussed incident wave (usually a Gaussian mode) limits what is experimentally possible, but experiments with single atoms [16] and molecules [10] using solid immersion lenses for tight focussing are yielding encouraging results, showing large absorption along with a significant phase shift. By using a multi-level system exhibiting electromagnetically induced transparency effects, the optical properties of the single atom can be dynamically controlled, opening the door to optical switching using single atoms and photons [17], the so-called 'optical transistor' that can be used in quantum information processing [18, 19].

In other fields, tremendous progress has been made with single emitters in the mesoscopic regime using gold [20] and silver [21] nanospheres, which have great potential in biological imaging systems, because surface plasmon resonances enhance their absorption, a great advantage over organic dyes [22, 23]. One approach to measuring the optical properties of these nanosphere systems involves interferometry exploiting an effective index approach [24].

The Maxwell-Garnet (MG) theory [25] that is employed in these experiments may help us answer the question of when a collection of emitters becomes a bulk medium. In the MG theory, a filling factor (either volume or surface, dependent on geometry) determines the behaviour of the system. Using nonresonant light, it has been suggested that the collection of emitters can be considered an effective medium (and thus conform to MG theory) when the individual nanospheres are separated by at most $\lambda_{\mathrm{eff}}/2\pi$ [26] (where $\lambda_{\mathrm{eff}} = \lambda/2\pi$), corresponding to a surface filling factor of around 1 % in their experiment.

We can estimate the filling factor for our resonant atomic system. In the dilute limit, the resonant optical cross-section per atom is given by

$$\sigma_0 = 3\lambda^2/2\pi, \tag{6.4}$$

which we now convert into a volume, via $r = \sqrt{\sigma_0/\pi}$, yielding

$$V_{\mathrm{atom}} = \frac{4\pi}{3}r^3 = \frac{\sqrt{6}}{\pi^2}\lambda^3. \tag{6.5}$$

We have a (volume) atomic number density N. Then the volume filling factor is given simply by

$$f_V = N V_{\mathrm{atom}}. \tag{6.6}$$

For the Rb D-lines, $\lambda \sim 8 \times 10^{-7}$ m, which gives $\sigma_0 \sim 3 \times 10^{-13}$ m^2. Using the effective index approach, and assuming that a filling factor of at least 10 % is needed (an overestimate, based on the nanosphere experiments), we have the condition

$$N > 0.1/V_{\mathrm{atom}} \tag{6.7}$$
$$\gtrsim 10^{11} \mathrm{cm}^{-3}, \tag{6.8}$$

which is fulfilled for temperatures $T \gtrsim 50\,^{\circ}$C. Even assuming some orders of magnitude of error in the estimation, the effective index approach is valid in our experiment.

This approach also predicts that we can achieve filling factors $f_V > 1$, where the separation becomes smaller than r, for temperatures $T \gtrsim 150\,^{\circ}$C. In this case, the dipole–dipole interaction starts to become important, reducing the effective optical cross-section per atom (as discussed in Sect. 5.4.2).

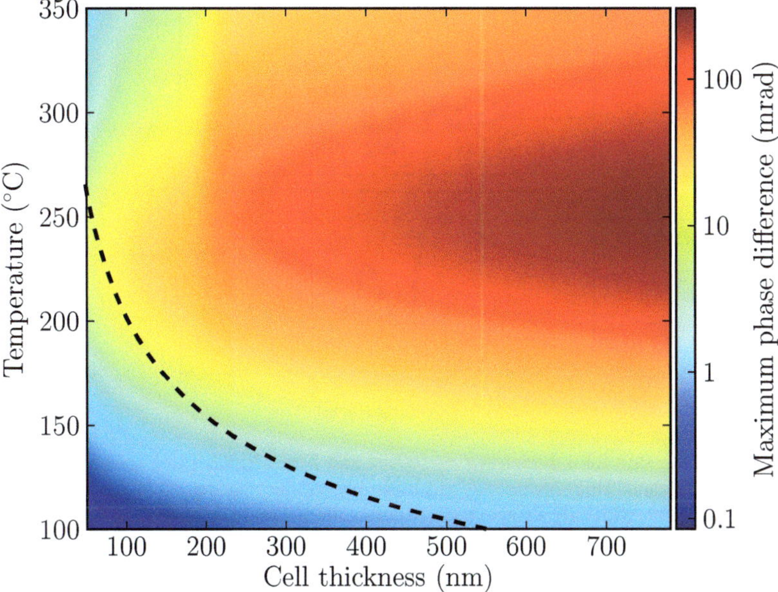

Fig. 6.6 Theoretical maximum phase shift over the line as a function of temperature (density) and cell length, for $\Delta_{AOM} = 160\,\mathrm{MHz}$. The sharp change at $\ell = \lambda/4 = 195\,\mathrm{nm}$ is due to the change in broadening coefficient, detailed in the previous chapter. The dashed line represents the temperature where the cell thickness is equal to the mean interatomic spacing $\ell = r_{av}$

6.4.1 Experiment

By tuning the density and thickness of the vapour to a region where the mean interatomic spacing r_{av} is equal to the vapour thickness the beam interacts with, on average, only a single layer of atoms. As shown in Fig. 6.6, there is a range of experimental parameters where this condition is met (dashed line). This allows us to tune the atomic interactions while still only probing a single layer. The main limitation is due to experimental signal-to-noise. Using the susceptibility model we can predict where the maximum signal occurs, as shown in Fig. 6.6. The maximum relative phase shift for a single atomic layer occurs around $\ell = 100\,\mathrm{nm}$ and $T = 220\,^\circ\mathrm{C}$, and in Fig. 6.7 we show the experimentally measured phase shift (a) and inferred index (b) for these conditions. Despite the layer being only a single atom thick, on average, we still observe a measurable phase shift commensurate with a maximum refractive index $n \sim 1.025$, and the data still fit exceptionally well to theory in this extreme limit, a clear indication that the vapour can still be considered as a medium, rather than individual atoms.

What is also apparent from Fig. 6.7 is the excellent signal-to-noise ratio; the experimental technique gives us a sensitivity to phase shifts of the order of 1 mrad, which could be further improved by increasing the AOM driving frequency, thereby creating a bigger difference in refractive index between the two beams.

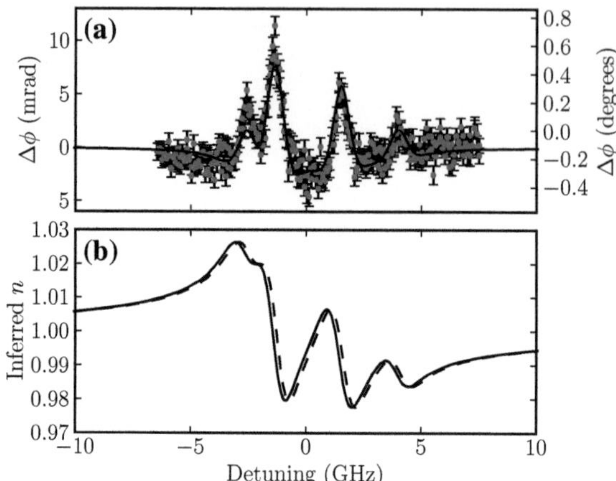

Fig. 6.7 **a** Phase shift for a thickness $\ell = 100$ nm and $T = 220\,^{\circ}$C. Three separate experimental data sets are overlaid. This corresponds to a mean interatomic spacing $r_{av} = \ell$, so on average the laser beam interacts with only a single layer of atoms. Remarkably there is still a measurable phase shift which corresponds to a significant refractive index for the layer (**b**)

6.5 Outlook

We have adapted the heterodyne interferometric technique previously applied to single molecules across a single, narrow resonance to vapors where there are many resonances which are generally broader than the frequency splitting of the two beams. Given a high enough $N\ell$ product, which approximately determines the magnitude of the phase shift, and providing one has a suitable model for the susceptibility, there is no reason why this sensitive technique cannot be applied to a variety of other systems where accurate phase information is required. The technique is readily applicable to longer vapours, including reference cells, with minimal modification to the setup—only the lenses need be removed to avoid optical pumping effects.

In addition, heterodyne interferometry has some advantages over transmission spectroscopy, the most important being that it is insensitive to changes due to the varying cell reflectivity, as only the relative phase information is important. Furthermore, there is no additional calibration of the raw data. Unlike transmission spectroscopy, the only calibration is the frequency axis, which is set by the wavemeter, which can be calibrated independently. The main disadvantage is the requirement that the optical depth not be too high, as this adversely affects the signal-to-noise ratio. For small optical depths, this method could provide another technique for analysing line widths and shifts, as the same susceptibility model is used.

In this chapter, we have shown that the refractive index of a Rb vapour on resonance can be as large as 1.31 in theory, and experimentally measured an index of 1.26 ± 0.02. To the best of the author's knowledge, this measurement is the highest refractive index measured in an atomic vapour.

By tuning the density and thickness of the layer, it is possible to investigate a vapour that is on average only one atom thick. Remarkably, even under these conditions the data fits well to theory assuming a medium with a well-defined thickness, and we observe phase shifts on the order of a few mrad.

One feature common to both the refractive index plots of Figs. 6.5 and 6.7 is the region of anomalous dispersion ($dn/d\omega < 0$) on resonance. This gives rise to a negative group velocity, i.e. the velocity at which the peak of a light pulse propagates through the medium, and causes the peak of an optical pulse to travel more quickly through the medium than through vacuum. This superluminal propagation will be the topic of the next chapter.

References

1. J.B. Pendry, A.J. Holden, D.J. Robbins, W.J. Stewart, Magnetism from conductors and enhanced nonlinear phenomena. IEEE Trans. Microw. Theor. Tech. **47**, 2075 (1999)
2. D. Smith, W. Padilla, D. Vier, S. Nemat-Nasser, S. Schultz, Composite medium with simultaneously negative permeability and permittivity. Phys. Rev. Lett. **84**, 4184 (2000)
3. M. Choi et al., A terahertz metamaterial with unnaturally high refractive index. Nature **470**, 369 (2011)
4. G. Dolling, M. Wegener, C.M. Soukoulis, S. Linden, Negative-index metamaterial at 780 nm wavelength. Opt. Lett. **32**, 53 (2007)
5. D. Schurig et al., Metamaterial electromagnetic cloak at microwave frequencies. Science **314**, 977 (2006)
6. W. Cai, U.K. Chettiar, A.V. Kildishev, V.M. Shalaev, Optical cloaking with metamaterials. Nat. Photonics **1**, 224 (2007)
7. T. Ergin, N. Stenger, P. Brenner, J.B. Pendry, M. Wegener, Three-dimensional invisibility cloak at optical wavelengths. Science **328**, 337 (2010)
8. J.C. Owens, Optical refractive index of air: dependence on pressure, temperature and composition. Appl. Opt. **6**, 51 (1967)
9. A.S. Zibrov et al., Experimental demonstration of enhanced index of refraction via quantum coherence in Rb. Phys. Rev. Lett. **76**, 3935 (1996)
10. M. Pototschnig et al., Controlling the phase of a light beam with a single molecule. Phys. Rev. Lett. **107**, 063001 (2011)
11. M. Fleischhauer et al., Resonantly enhanced refractive index without absorption via atomic coherence. Phys. Rev. A **46**, 1468 (1992)
12. J. Keaveney et al., Optical transmission through a dipolar layer, arxiv.org, 1109.3669v2 (2011), arXiv:1109.3669v2
13. C. O'Brien, P.M. Anisimov, Y. Rostovtsev, O. Kocharovskaya, Coherent control of refractive index in far-detuned Λ systems. Phys. Rev. A **84**, 063835 (2011)
14. Z.J. Simmons, N.A. Proite, J. Miles, D.E. Sikes, D.D. Yavuz, Refractive index enhancement with vanishing absorption in short, high-density vapor cells. Phys. Rev. A **85**, 053810 (2012)
15. G. Zumofen, N. Mojarad, V. Sandoghdar, M. Agio, Perfect reflection of light by an oscillating dipole. Phys. Rev. Lett. **101**, 180404 (2008)
16. S.A. Aljunid et al., Phase shift of a weak coherent beam induced by a single atom. Phys. Rev. Lett. **103**, 153601 (2009)
17. T. Kampschulte et al., Optical control of the refractive index of a single atom. Phys. Rev. Lett. **105**, 153603 (2010)
18. Q. Turchette, C. Hood, W. Lange, H. Mabuchi, H. Kimble, Measurement of conditional phase shifts for quantum logic. Phys. Rev. Lett. **75**, 4710 (1995)

19. J. Hwang et al., A single-molecule optical transistor. Nature **460**, 76 (2009)
20. S. Kubo et al., Tunability of the refractive index of gold nanoparticle dispersions. Nano Lett. **7**, 3418 (2007)
21. R. Brenier, Enhancement of light transmission through silver nanoparticles. J. Phys. Chem. C **116**, 5358 (2012)
22. E. Boisselier, D. Astruc, Gold nanoparticles in nanomedicine: preparations, imaging, diagnostics, therapies and toxicity. Chem. Soc. Rev. **38**, 1759 (2009)
23. D.A. Giljohann et al., Gold nanoparticles for biology and medicine. Angew. Chem. (International ed. in English) **49**, 3280 (2010)
24. G.H. Cross et al., The metrics of surface adsorbed small molecules on the Young's fringe dual-slab waveguide interferometer. J. Phys. D **37**, 74 (2004)
25. J.C. Maxwell Garnett, Colours in metal glasses and in metallic films. Phil. Trans. R. Soc. Lond. A **203**, 385 (1904)
26. G.H. Cross, Light under a matter field microscope, arxiv.org, 1303.1390 (2013)

Chapter 7
Fast Light in Dense Thermal Vapour

In the previous chapter we demonstrated a large refractive index enhancement in Rb vapour via a continuous-wave (CW) interferometry technique. In this chapter we further examine the extreme properties of this medium via pulsed excitation. We show that accompanying the large refractive index there is also a GHz-bandwidth region of very large anomalous dispersion, leading to the superluminal propagation of sub-nanosecond optical pulses.

7.1 Introduction

Controlling the speed of an optical pulse in a medium is a topic of much interest [1, 2], with potential applications in optical communications and quantum information processing [3–5]. The speed of the peak of an optical pulse is determined by the group velocity of the medium, and is given by

$$v_{\mathrm{g}} = \frac{\mathrm{d}\omega}{\mathrm{d}k}.$$ (7.1)

Using the relationship $k = n(\omega)k_0 = n(\omega)\omega/c$, where ω is the carrier (central) frequency of the pulse, with a frequency-dependent (real part of the) refractive index $n(\omega)$ in the medium, we find

$$\frac{\mathrm{d}k}{\mathrm{d}\omega} = \frac{1}{c}\left(n(\omega) + \omega\frac{\mathrm{d}n(\omega)}{\mathrm{d}\omega}\right).$$ (7.2)

The group velocity can then be expressed as

$$v_{\mathrm{g}} = \left(\frac{\mathrm{d}k}{\mathrm{d}\omega}\right)^{-1} = \frac{c}{n(\omega) + \omega\frac{\mathrm{d}n(\omega)}{\mathrm{d}\omega}} = \frac{c}{n_{\mathrm{g}}(\omega)},$$ (7.3)

J. Keaveney, *Collective Atom–Light Interactions in Dense Atomic Vapours*,
Springer Theses, DOI: 10.1007/978-3-319-07100-8_7,
© Springer International Publishing Switzerland 2014

where we have defined the group index n_{g}. As we have previously seen in Chaps. 2 and 6, around a resonance the refractive index varies significantly with frequency. The group velocity can therefore be positive or negative, and varies dramatically over a resonance where $dn/d\omega$ changes sign.

An optical pulse traveling through a resonant medium of length ℓ will therefore be delayed compared to a pulse travelling the same distance through vacuum. The delay (of the peak of the pulse) can be expressed as

$$\Delta t = \frac{\ell}{v_{\mathrm{g}}} - \frac{\ell}{c} = (n_{\mathrm{g}} - 1)\frac{\ell}{c}. \tag{7.4}$$

If $n_{\mathrm{g}} < 0$, the resonant pulse will therefore be advanced in time with respect to the vacuum pulse. 'Slow' light is therefore to be associated with a pulse delay, and 'fast' light with a pulse advance (grouping together cases where $v_{\mathrm{g}} > c$ and $v_{\mathrm{g}} < 0$), both of which will be dealt with separately in later sections.

7.1.1 Kramers-Kronig Relations

In an ideal situation, the pulse would be delayed or advanced without any change in its shape (distortion) or loss of amplitude (absorption). We have already shown that absorption is determined by the imaginary part of the electric susceptibility, and the refractive index profile is dependent on the real part of the atomic susceptibility (Sect. 2.2). The susceptibility of an ideal medium would thus combine a vanishing imaginary part with a finite real part.

The Kramers-Kronig relations [6, 7] express the fundamental link between the real and imaginary parts of the susceptibility. Mathematically, they can be expressed as [8]

$$\mathrm{Re}\,\chi(\omega_0) = \frac{2}{\pi}P\int_0^\infty \frac{\omega\,\mathrm{Im}\chi(\omega)}{\omega^2 - \omega_0^2}\,d\omega, \tag{7.5a}$$

$$\mathrm{Im}\,\chi(\omega_0) = -\frac{2}{\pi}P\int_0^\infty \frac{\omega_0\,\mathrm{Re}\chi(\omega)}{\omega^2 - \omega_0^2}\,d\omega, \tag{7.5b}$$

where P denotes the Cauchy principle value (see, for example, Chap. 14 of [9]). The real part of the susceptibility at one frequency depends on the whole frequency spectrum of the imaginary part, and vice-versa. In other words, one cannot be present without the other in close proximity. This has profound implications for pulse propagation in dispersive media, strictly constraining the possible behaviour.

In 'normal' media (no amplification) $\mathrm{Im}\,\chi(\omega) > 0$ for all ω [10]. Using this, it can be shown by differentiating equation (7.5a) that in the wings of an absorption line there is so-called 'normal' dispersion where $dn/d\omega > 0$, while on resonance we have so-called 'anomalous' dispersion with $dn/d\omega < 0$. We thus have regions of slow-light in the wings, while fast-light occurs on resonance.

It is possible to reverse this in gain media, where instead of attenuation (Im $\chi(\omega) > 0$), there is amplification (Im $\chi(\omega) < 0$). Gain occurs when there is a population inversion in the system, the simplest case being that of a two-level atom, so that $\rho_{22} > \rho_{11}$. The susceptibility from Eqs. (2.21) and (2.16) then changes sign from absorbing media—both the real and imaginary parts are affected, so that in gain media there is slow-light on resonance and fast-light in the wings.

7.1.2 Wave Packets, Bandwidth and Group Velocity Dispersion

A truly monochromatic wave is infinite in its extent, with an electric field that oscillates at a single (angular) frequency ω. The Fourier transform of the electric field, its frequency spectrum (or power spectrum, as it is sometimes called), is therefore a delta-function at ω. An optical pulse must be switched on and off at some point in time, and so cannot extend infinitely far. The frequency spectrum must then have some width, that depends on the pulse shape and duration. Intuitively, if we consider a very long 'pulse', e.g. a CW laser which is turned on and off at some point, then the electric field is very nearly monochromatic. The frequency spectrum (or power spectrum) is exactly the Fourier transform of the pulse. As the pulse gets shorter or sharper (e.g. a square wave) then the Fourier transform gets wider. One can also think of this in terms of the uncertainty principle; $\Delta E \Delta t \geq \hbar$, or $\Delta\omega\Delta t \geq 1$. The pulse has a finite temporal profile Δt, so it must have a corresponding frequency-space uncertainty $\Delta\omega$.

The frequency spectrum is a very useful tool for understanding optical pulses. Fourier analysis tells us we can decompose the electric field profile into a series of monochromatic waves, each with an amplitude $F(\omega)$. This is depicted in Fig. 7.1. These frequency components are monochromatic and infinite in extent, and it is therefore only where there is constructive interference between the components that a pulse is formed. The more components are added the more localised the region of constructive interference becomes; wider in frequency space means narrower in time (and real space), and vice-versa.

It is for this reason that pulses are often called 'wave packets'; they are literally a packet, or group, of waves that travel together. The concept of group velocity is therefore the speed at which the point of constructive interference moves. A dispersive medium can therefore modify this quantity, since each component frequency of the wave packet experiences a different refractive index, $n(\omega)$.

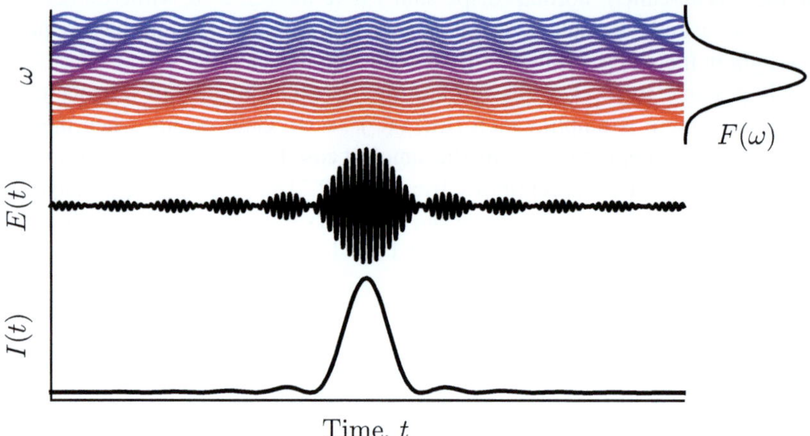

Fig. 7.1 Illustration of the Fourier composition of a wavepacket. An optical pulse with intensity $I(t)$ and electric field $E(t)$ is localised in time, but composed of many monochromatic Fourier components, with relative contribution given by the Fourier transform $F(\omega)$. The individual components are infinite in time (and therefore space) but only add in phase at localised points. The propagation of the pulse is thus determined by the propagation of each of the Fourier components

7.1.2.1 Bandwidth

The intensity profile of the experimental pulses used in this thesis approximate well to Gaussians, so we have

$$I(t) = I_0 \, e^{-(t-t_0)^2/\tau^2},\tag{7.6}$$

with the full-width at half-maximum (FWHM) $t_{\mathrm{FW}} = 2\sqrt{\ln 2}\,\tau$. The Fourier transform of a Gaussian is another Gaussian, so the pulse spectral profile $F(\nu)$ has the same form as the electric field profile, with a FWHM ν_{FW}. The full-widths are related by

$$t_{\mathrm{FW}}\nu_{\mathrm{FW}} = 0.441.\tag{7.7}$$

We now define the pulse bandwidth as

$$\nu_{\mathrm{BW}} = \omega_{\mathrm{BW}}/2\pi = 2\nu_{\mathrm{FW}},\tag{7.8}$$

which is twice the FWHM. The frequencies contained in this range make up over 98 % of the pulse power spectrum. For the pulses used in this experiment, $t_{\mathrm{FW}} = 800$ ps which corresponds to $\nu_{\mathrm{FW}} \approx 550$ MHz and $\nu_{\mathrm{BW}} \approx 1.1$ GHz.

7.1.2.2 Group Velocity Dispersion

The group velocity dispersion (GVD) is a measure of how uniform the group velocity is over the entire pulse bandwidth and is defined as $dv_{\rm g}/d\omega$. If the whole of the pulse profile (in frequency space) is in a flat region of GVD, then all frequency components of the pulse travel at the same speed and therefore the pulse will experience no distortion. If this is not the case, then the pulse temporal profile will be distorted. This can be exploited to compress or narrow [11], stretch [12, 13] and split a single pulse into many [14]. Birefringence can also be harnessed to separate, for example, the left and right circular polarisation components of linearly polarised light pulse [13].

7.2 Controlling the Group Velocity

7.2.1 Slow Light

An important feature of the atomic susceptibility is how the real and imaginary parts scale with the detuning from resonance Δ. If we consider the simple two-level atom from Chap. 2, the real and imaginary parts of Eq. (2.22) yield

$$\mathrm{Re}\,\chi = -\frac{\Delta}{(\Gamma^2/4) + \Delta^2}, \tag{7.9a}$$

$$\mathrm{Im}\,\chi = \frac{\Gamma}{(\Gamma^2/4) + \Delta^2}. \tag{7.9b}$$

Far from resonance, absorption reduces as $1/\Delta^2$ whereas the dispersive component scales as $1/\Delta$. It can therefore be advantageous in slow-light experiments to work in the far wings of a resonance [15] or between two resonance lines, where absorption is negligible but dispersive effects are still significant. Furthermore, because the profiles at large detuning are gently sloping, the GVD is nearly zero over a large frequency range. In thermal atomic vapours, using this principle it has been possible to achieve significant fractional pulse delays, $\Delta t/t_{\rm FW}$, of nanosecond pulses with relatively little absorption [13], with very little pulse distortion, as Doppler broadening provides a GHz-bandwidth medium. The delay using this technique can be extremely large, fractional delays up to 80 have been observed with 740 ps pulses (59 ns delay) in caesium vapour [16] with a corresponding group index $n_{\rm g} \approx 200$.

The highest group indices, however, are realised using a resonant interaction. Nearer to the resonance line, the refractive index varies more rapidly and so the group index is larger. However, this comes at the cost of rapidly increasing optical depth. To circumvent the optical depth problem, many slow-light experiments use electromagnetically induced transparency (EIT) [17–19]. For a comprehensive review of EIT and its applications, see the review by Fleischhauer [20].

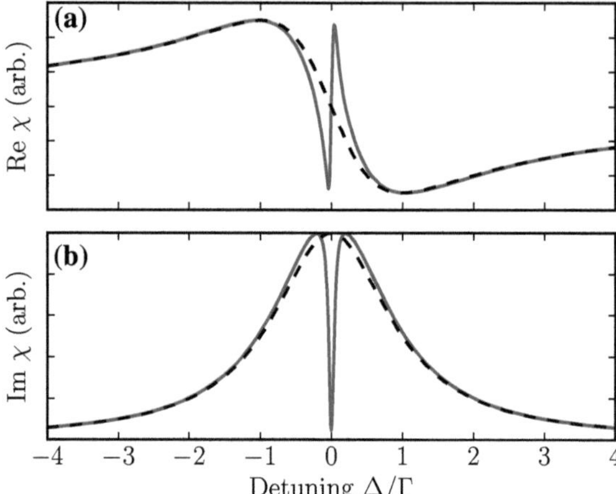

Fig. 7.2 Real and imaginary parts of the susceptibility around a resonance in an EIT-type three-level system (*red*), compared with the two level equivalent (*black dashed*). The steep gradient in Re χ around $\Delta = 0$ coincides with a transparency window in the absorption profile. This creates a near-transparent slow-light region of the type used in many experiments. Calculations based on the weak probe susceptibility, Eq. (5) in reference [21], neglecting Doppler broadening (colour in online)

EIT is a nonlinear optical process which typically involves two photons that resonantly couple three atomic energy levels. An example susceptibility profile is shown in Fig. 7.2, as a function of the probe laser detuning. Without the coupling laser (black dashed lines), the system reduces to the two level case presented in Chap. 2. When the coupling laser is on, destructive interference between excitation pathways reduces the resonant absorption of the (weak) probe laser when the (strong) coupling laser is resonant with the second transition, as is the case for Fig. 7.2. The interference profoundly changes the refractive index profile, so that on resonance there is now a region of steep normal dispersion corresponding to a large positive group index, where before there was anomalous dispersion. This coincides with a reduction of the imaginary part of the susceptibility.

Coherence between the first and third levels that are not directly coupled by the optical fields is generated when the two-photon detuning is zero; neither laser needs to be exactly resonant with the second (intermediate) state. The decay rate of this coherence sets the linewidth of the EIT feature, which is a contributing factor to the magnitude of the slow-light effect. It is therefore prudent to arrange for the first and third states to be ground-states of the system, which have very long radiative lifetimes, so nearly all slow-light experiments based on EIT use this so-called Λ-type excitation scheme. The coherence decay rate will then be limited by other dephasing processes. In thermal vapours, time-of-flight broadening is the most important (the excitation is essentially Doppler-free as it is done with copropagating beams with the same frequency to 1 part in 10^5), and this limits the linewidth to around 30 kHz [22]. Using paraffin-coated walls which introduce coherence-preserving collisions [23], and buffer gases which increase the interaction time with the beam,[1] it is possible to reduce the linewidth further. Linewidths below 30 Hz have been achieved with Rb buffered by neon gas [25]. Using this approach, with a sub-kHz linewidth in neon-buffered Rb, Kash et al. [26] observed a group velocity of just 90 m/s. A similar reduction in group velocity was also observed by Budker et al using a nonlinear magneto-optic effect [27].

By moving to the ultra-cold regime, however, one can further reduce the dephasing, and with a Bose-Einstein condensate (BEC) it is possible to do this with a relatively large atomic density. In a landmark experiment, Hau et al. [28] observed a reduction of the group velocity to just 17 m/s, corresponding to a group index $n_g > 2 \times 10^7$, using a BEC of sodium atoms at temperatures of $\sim$100 nK, with an atomic number density of 8×10^{13} cm^{-3}.

7.2.1.1 Stored Light

It can be shown[2] that the group velocity depends on the intensity of the applied coupling field. By dynamically tuning the coupling Rabi frequency it is possible to reduce the group velocity to zero, storing the light in the medium [29, 30]. The pulse energy propagates in the medium as a 'dark state polariton', a superposition state of atomic coherence and electromagnetic field. The pulse can therefore be stored by making the polariton purely atomic, and read out again at a later time by converting back to a purely electromagnetic component. This storage and retrieval was first demonstrated by Liu et al. [31] and Phillips et al. [32] in quick succession, where the optical pulse was stored for many times the pulse temporal width before readout.

Photon storage and retrieval continues to be a topic of great interest [33], and has been performed in ultracold atomic systems with storage times up to 1 min [34]. In similar ultracold experiments at the few-photon level with long-lived Rydberg states, control over the quantum state of the stored photons has been demonstrated [35] via microwave coupling between different Rydberg states.

---

[1] The buffer gas introduces some additional dephasing due to collisions, but this is largely negligible for ground state atoms. For excited state atoms the collisional dephasing is more important, leading to a broadening of the EIT feature [24].

[2] The group velocity can be derived from the susceptibility of the three-level system. In the weak probe limit, and when both lasers are resonant, the group index and velocity can be expressed as [20, 21, 26, 28]

$$n_g(\Delta_1 = 0) = n + \omega_p \frac{dn}{d\omega_p} \approx n + \omega_p \frac{1}{2} \frac{\partial \chi}{\partial \Delta_1}\bigg|_{\Delta_1 = \Delta_2 = 0}$$

$$v_g(\Delta_1 = 0) = \frac{\hbar c \epsilon_0}{2\omega_p} \frac{1}{N d_{12}^2} \frac{(\gamma_{21}\gamma_{31} + \Omega_c^2/4)^2}{\Omega_c^2/4 - \gamma_{31}^2},$$

where ω_p is the probe detuning, N is the atomic density, Ω_c is the coupling Rabi frequency and γ_{21} and γ_{31} are the decay rates of the $|2\rangle - |1\rangle$ and $|3\rangle - |1\rangle$ coherences, respectively. The coupling Rabi frequency is proportional to the square root of the intensity, $I_c \propto \Omega_c^2$ of the coupling laser.

7.2.2 *Fast Light, Causality and Information Velocity*

The concept of fast light at first glance seems to be at odds with Einstein's special theory of relativity [36], leading to causality-violating implications, i.e.that the energy, and therefore information, in the pulse travels faster than c. There has been a great deal of literature on the subject of information velocity in fast light media, starting with discussions between Sommerfeld and Brillouin [37]. A brief summary is given here, but the interested reader should consult, for example, references [38–41] for a more in depth discussion and analysis.

In order to transmit information, something must vary in time. If we take the simplest case of this, a binary state where the signal is either on or off, the only time information is transmitted is when the signal changes state. A completely monochromatic wave, being infinite in its extent, can therefore not carry any information. Hence a phase velocity larger than c does not imply information velocity. To transmit information, we need a signal that varies in time, or in other words, a wavepacket.

One might think, then, that the speed of information flows with the group velocity of the wavepacket. For a slow-light medium, this can often be the case, as it can be understood in terms of an optical pulse being a physical object that is slowed down. The same cannot be said for fast light, however, where the group velocity can be greater than c or even negative, which would imply either non-causal behaviour, or that the signal propagates backwards in space (and hence time).

To understand information flow in fast light media, we need to further refine the definition of 'information'. If we have an optical pulse that is, say, a Gaussian function, then in principle the pulse is analytic over its whole spatial extent. In a naive approximation, the detection of the pulse at any time is limited only by some detector efficiency. Once again then we return to the picture of a spatially and therefore temporally infinite waveform, and the time between generating the pulse and detecting it is in principle only dependent on the detector.

The only way to produce a true localised signal in time or space that cannot be predicted by having a better detector is to have a discontinuous, or non-analytic, jump in the value of the signal at some time. If we return to the Fourier picture of Fig. 7.1, it is clear that the only way to do this is to have an infinitely large signal bandwidth. In a physical system the anomalous dispersion curve on which the pulse is centred must be necessarily finite in magnitude and width, so that the non-analytic pulse *has to be* distorted. The distortion causes 'optical precursors' to appear, which are the signals associated with the zero (dc) and infinite frequency Fourier components [42], which travel at a *maximum* velocity of c. Thus the information can only travel at a speed that is less than or equal to c.

On a more practical level, Kuzmich et al. [38] showed that quantum fluctuations (noise) limit the detection of any real signal, again coming to the conclusion that due to these fluctuations the point at which one can discern a signal is pushed back in time, reducing the signal velocity below c. In terms of signal-to-noise ratio (SNR), the vacuum pulse always has a higher SNR and therefore it can always be detected first.

7.2.3 *Experimental Observations of Fast Light*

Until recently, most experimental studies of fast light media have used Gaussian input pulses, as these are relatively easy to produce, and analytically simple. The theory of their propagation through dispersive media was first presented by Garrett and McCumber [42] and Crisp [43], who showed that the usual group velocity equations still hold for fast light provided there is no distortion of the pulse. Their theoretical predictions were realised a decade later by Chu and Wong [44].

The classic experimental demonstration of fast light in gain media came at the start of this century, in a paper by Wang et al. [45]. Using the anomalous dispersion region located between two gain lines in optically pumped atomic Cs, they demonstrated superluminal propagation of a microsecond-duration optical pulse, with a group index $n_g = -310$. Similar experiments from Akulshin et al. followed [46], achieving larger (negative) indices ($n_g = -5,100$). Though the group indices demonstrated in these two experiments implies backward propagation of the pulse in the medium, an explicit demonstration of backward propagation was first presented using a fiber amplifier with a dip in the gain profile [47], with a measured index of $n_g = -4,000$.

7.3 Group Refractive Index in Dense Thermal Rb Vapour

Methods based on gain media for fast light and EIT for slow light share a common limitation: because the absolute change in index Δn is small, in order for a large (positive or negative) group index, the frequency range over which the index changes, $\Delta \omega$ must also be very small. They are therefore limited to low bandwidth applications, with relatively long optical pulses.

As we have seen in Chap. 6, the refractive index of dense Rb vapour can be very large, so Δn is of the order of 10^{-1}. The cost of this is a broad resonance, due to the atomic interactions. The group index, which can be roughly expressed as $\omega \Delta n / \Delta \omega$, can therefore still be very large, with an absolute value larger than 10^5 as we will see later in the section. An order-of-magnitude comparison of realistic experimental parameters for different slow- and fast-light methods is presented in Table 7.1.

For high density thermal vapours probed with resonant laser light, optical depth is the limiting factor. By using nano-cells however, this can be partially circumvented, though obviously at the cost of fractional delay/advance. The main advantage of using thermal media is the GHz-bandwidth provided by the broad atomic lines, due to the dipole–dipole interactions between atoms. The large bandwidth means that relatively short pulses can be used, down to around a nanosecond, without significant pulse distortion. In addition, by tuning the laser frequency over several GHz it is possible to move from a fast-light to a slow-light medium, and by tuning the density, and therefore the linewidth of the vapour, the magnitude of the slow- or fast-light effect can be modified.

Table 7.1 Order-of-magnitude comparison of slow- and fast-light methods

	Δn	$\Delta\omega$	n_g	ℓ	Δt	$\Delta\omega\Delta t$
Slow light						
Off resonance, thermal	10^{-3}	10^{9}	100	10^{-1}	10^{-7}	10^{2}
EIT, ultracold	10^{-2}	10^{6}	10^{7}	10^{-4}	10^{-5}	10
Near-resonance, thin cells	-10^{-1}	10^{9}	10^{4}	10^{-7}	10^{-10}	0.01
Fast light						
Gain media, thermal	-10^{-6}	10^{6}	-10^{2}	10^{-2}	10^{-8}	0.01
Resonant, thin cells	-10^{-1}	10^{9}	-10^{5}	10^{-7}	10^{-10}	0.1

Group index $n_g \sim \omega\Delta n/\Delta\omega$ with $\omega/2\pi \sim 3\times10^{14}$ Hz for optical frequencies. $\Delta\omega\Delta t$ is the fractional delay/advance of the pulse, assuming $\Delta\omega \sim 1/t_{\text{FW}}$. Calculations are based on the experimental results in [13, 25, 26, 28, 45–47]

To calculate the group index in the medium, we use the susceptibility model developed in Chaps. 2 and 3. We plot this as a function of vapour temperature and detuning in Fig. 7.3. Unlike the refractive index (see Fig. 6.4), the optimum group index does not coincide with the highest temperature; instead we predict the largest group index to occur at a temperature $T \approx 255\,^\circ\text{C}$ on resonance with the strongest hyperfine transition in ^{85}Rb, from $F_g = 3$ to $F_e = 4$, which occurs at a detuning $\Delta/2\pi = -1.2$ GHz (we omit the line shift from the figure for clarity, since it does not significantly affect the position of the maximum). In the pulse propagation experiment presented later in this chapter, this detuning is used as the carrier detuning for the optical pulse. This maximum occurs because the susceptibility, and therefore the refractive index, up to this temperature increases more in magnitude than width. As temperature is increased further, the width increases more than the magnitude and therefore $dn/d\omega$ decreases.

In Fig. 7.4, we plot the refractive index and group index profiles for the optimal case, as a function of the detuning from resonance. The dashed line shows the carrier detuning of the optical pulse used in the experiment, while the red shaded area represents the pulse bandwidth. The opacity of the shading represents the power spectrum of the pulse. Over the central region ($\omega_c \pm \omega_{\text{FW}}$) of the pulse, the group index varies by a factor of 1.5 from the peak value of -1.20×10^{5}.

7.4 Pulse Propagation

In Sect. 7.1.2 we discussed the construction of optical pulses from their Fourier components. The propagation of the pulse is therefore determined by the propagation of each of the individual Fourier components.

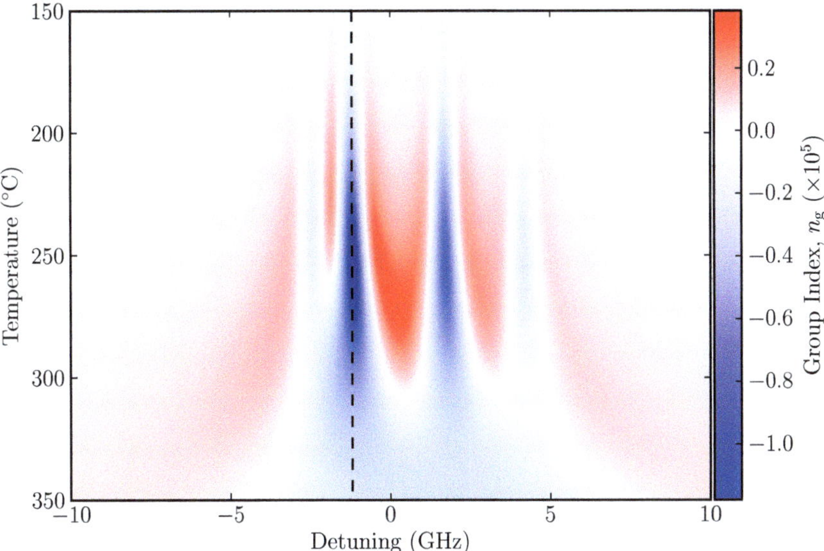

Fig. 7.3 Calculated group index as a function of temperature and linear detuning $\Delta/2\pi$ on the D2 line. The *dashed line* represents the carrier detuning for the pulse used in the experiment

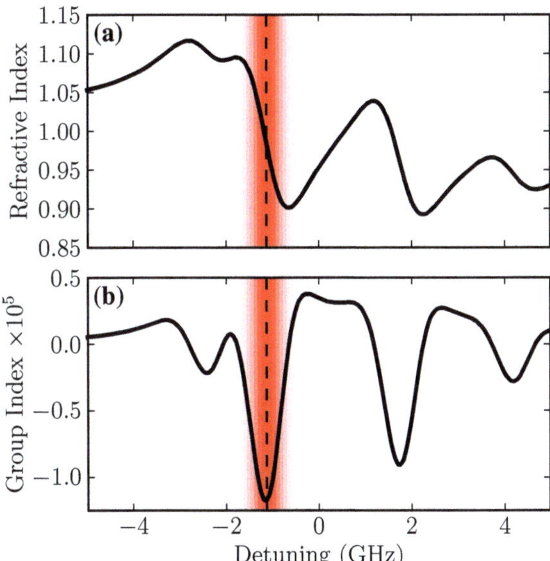

Fig. 7.4 Calculated refractive (**a**) and group (**b**) indices as a function of linear detuning $\Delta/2\pi$ at a temperature $T = 255\,°C$. The pulse centre frequency and bandwidth are denoted by the *dashed line* and *shaded area*, respectively

To begin, consider an input optical pulse with a Gaussian temporal profile. In free space the electric field envelope is given by

$$E(t) = E_0\, e^{-t^2/2\tau^2}.$$

(7.10)

We can then consider the plane-wave Fourier components

$$F(\omega) = \frac{1}{\sqrt{2\pi}} \int E(t)\, e^{i\omega t}\, dt.$$

(7.11)

Since we have a frequency dependent refractive index $n(\omega)$, the individual frequency components of the pulse each experience a different refractive index and thus a different phase shift. At the exit of the medium

$$F(z, \omega) = F(z = 0, \omega) e^{in(\omega)k_0 z}.$$

(7.12)

Since the refractive index is complex, the real part of the index gives rise to a phase shift $n_R k_0 z$, while the imaginary part attenuates $(n_I > 0)$ or amplifies $(n_I < 0)$ the component parts. To find the electric field profile at the exit of the medium, we simply take the inverse Fourier transform of Eq. (7.12)

$$E(z, t) = \frac{1}{\sqrt{2\pi}} \int F(z, \omega) e^{-i\omega t}\, d\omega.$$

(7.13)

Finally, we convert the field into an intensity, since this is what we measure experimentally, via

$$I(z, t) = \frac{\epsilon_0 c}{2}\, |E(z, t)|^2,$$

(7.14)

assuming the refractive index of air ≈ 1.

7.4.1 Subluminal Propagation

As a test of the model, Fig. 7.5 shows the pulse propagation code applied to an off-resonant slow-light medium. The experiment is the same as that presented in reference [13], and uses a 75 mm thickness vapour cell filled with 99 % ^{87}Rb, 1 % ^{85}Rb. The transmission and refractive index profiles over the D1 resonance line at a temperature $T = 100\,^\circ$C are shown in the inset. The optical pulse has a width $t_{\text{FW}} = 1$ ns and carrier frequency centred between the hyperfine ground state transitions ($\Delta_c/2\pi = 300$ MHz, dashed line in the inset), where $dn/d\omega > 0$. In the same way as for Figs. 6.4 and 7.3, the variation in the refractive index, and thus the group index, depends on the atomic density so the pulse experiences a slow-light medium and is delayed by an amount that varies with temperature. The theoretical

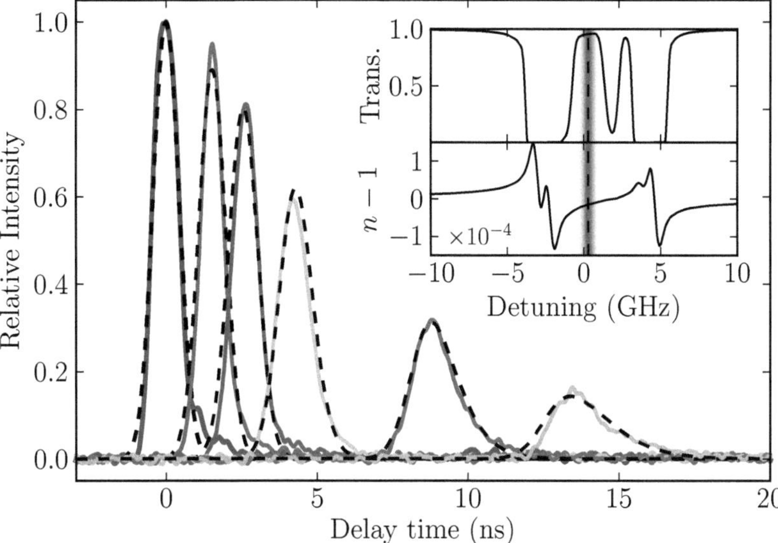

Fig. 7.5 Slow light: experiment and theory. Data (*solid lines*) and theory (*dashed lines*) showing pulse delay as a function of temperature in a 75 mm thickness vapour cell filled with 99 % ^{87}Rb. The temperatures are, left to right, {20, 105, 113, 123, 136, 144} °C. The inset shows the transmission and refractive index profiles at 100 °C. The pulse centre frequency and bandwidth are denoted by the *dashed line* and *shaded area*, respectively. Experimental data provided by Lee Weller

model fits well to the experimental data, with slight modification to the measured temperatures (less than 3 °C) that is within reasonable experimental uncertainty. There are no other adjustable parameters. Importantly the increase in the pulse width at the highest temperatures that arises due to finite group velocity dispersion is correctly predicted by the theory.

7.4.2 *Superluminal Propagation of Sub–Nanosecond Optical Pulses*

In the present experiment, we use a Pockels cell located between two crossed Glan-Taylor polarisers which rapidly rotates the polarisation of light, producing optical pulses with $t_{FW} = 800$ ps. The pulses are sent through the cell, focussed as in the previous experiments to minimise the thickness variation, and we measure the arrival time of photons on a single photon counting module (SPCM). A histogram can then be made of the arrival time of these photons over a period of some hours, which gives good signal-to-noise ratio; the noise is at the 1 % level, primarily limited by background light. See Appendix C for details of the data acquisition process.

Because of the time taken to collect data the laser needs to be stabilised, which we do electronically using a wavelength meter (stable to less than 10 MHz) at an arbitrary

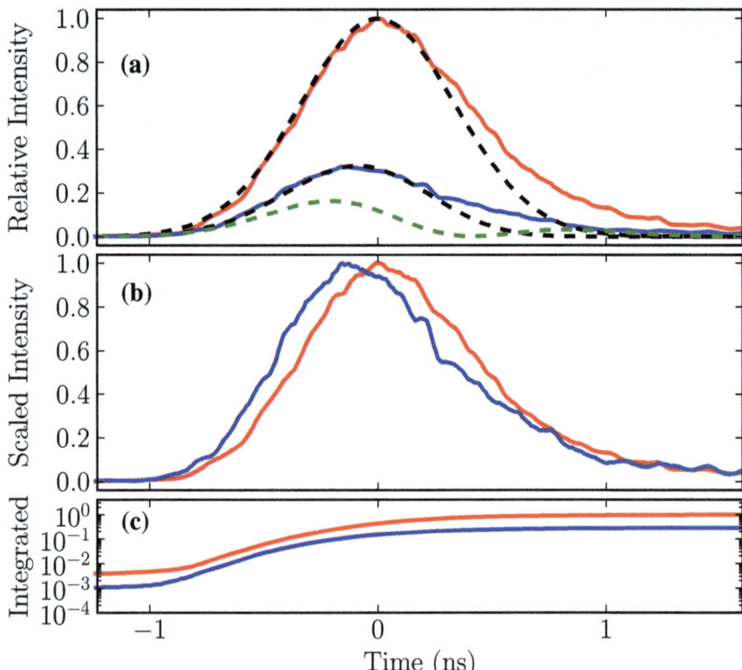

Fig. 7.6 (**a, b**) Superluminal propagation of a 800 ps pulse through a vapour thickness $\ell = \lambda/2$ and temperature $T = 255\,°C$. *Red* and *blue solid lines* are experiment, *dashed black lines* are theory with a 800 ps FWHM Gaussian input pulse. Off resonance ($\Delta_c/2\pi = -13$ GHz, *red line*) there is no interaction at this temperature and the pulse propagates through the vapour as it would through vacuum. On resonance, $\Delta_c/2\pi = -1.2$ GHz, the pulse is attenuated and arrives (0.13 ± 0.01 ns) earlier than the off-resonance reference pulse, corresponding to a group index $n_g = -(1.0 \pm 0.1) \times 10^5$. The *dashed green line* is the theory without dipole–dipole interactions. (**c**) Total integrated counts for both signals verifies preservation of causality—the probability of detecting a photon is always higher in the reference pulse

frequency. The optical power level is very low; the peak power in the pulse is of the order of 50 pW (with a 1 kHz repetition rate) so the excitation can be considered weak. In principle the power could be increased to speed up data acquisition, but the response of the SPCM modules used (Perkin Elmer SPCM-AQR-14-FC) has been shown to be non-linear for high count rates, skewing the distribution of arrival times.

Figure 7.6 shows experimental data for the theoretical optimum conditions, with a temperature $T = 255\,°C$ and cell thickness $\ell = \lambda/2$. We record a reference pulse where the laser is far detuned from resonance (carrier detuning $\Delta_c/2\pi = -13$ GHz) and there is no interaction with the atomic medium ($n_g \approx n \approx 1$). We then switch the laser frequency onto resonance to measure the effect of the medium.

On resonance we see attenuation to 0.35 of the maximum relative intensity and an advance of the peak by (0.13 ± 0.01) ns, corresponding to a large negative group index $n_g = (-1.0 \pm 0.1) \times 10^5$. The experimental pulse shape fits a Gaussian well on the leading edge, but not on the trailing edge. Because of this, the advance of the peak and its error were determined by fitting a Gaussian to the first half of both the reference and fast pulses and extracting a peak centre from this. The error in group velocity is propagated through from the error in peak centres.

Because of the large spectral bandwidth available, we see very little distortion of the advanced pulse, except on the trailing edge where resonance fluorescence causes the resonant pulse to have a slightly longer tail compared to the reference pulse (see Chap. 8). The dashed lines in Fig. 7.6a are theory curves based on the pulse propagation model, assuming an input pulse width of 800 ps. There is no fitting involved.

It is important to note that without the dipole–dipole interactions, the propagation of such temporally short pulses would not be possible without heavy distortion of the pulse shape. To illustrate this, the green dashed line in panel (a) shows the calculated pulse profile if there were no dipole-dipole interactions in the medium. In this case the pulse has a much larger bandwidth than the transition (spanning multiple excited state hyperfine levels) and therefore distortion of the output pulse is evident. Even though there is a superluminal component which is more advanced than for the interacting ensemble, there is also a sub-luminal component which can be seen as a small peak at $t \sim 0.8$ ns.

Panel (c) of Fig. 7.6 shows the total integrated counts over the detection period. The detection starts at $t = -15$ ns, which is why the two curves start at different points on the plot. The integrated counts verify that the advanced pulse preserves causality [38], since there is always a greater probability of detecting a photon in the reference pulse than the advanced pulse. From this we can immediately surmise that the information velocity of the advanced pulse must be less than c.

7.4.2.1 Fractional Advance

Since the advance scales linearly with the thickness of the medium, by making the vapour thicker we could increase the advance of the resonant pulse, but at the cost of additional attenuation, which scales exponentially with the thickness, as shown in Fig. 7.7.

In principle the maximum observable fractional advance is limited to around 2 by a signal-to-noise argument [48]: in attenuating media the signal is absorbed, so the point at which one can identify a 'signal' moves backward in time, effectively reducing the signal velocity. In gain media the situation is slightly different as the signal does not get attenuated. Instead, the fundamental quantum mechanical principle that amplification always introduces additional noise (see reference [40]) becomes important, then the same argument of identification of signal above the noisy background limits the observable advance. An analytic treatment by Macke et al. [49] reached similar conclusions.

Experimentally, fractional advances up to 0.25 have been observed [50], but with severe attenuation (2 % transmission). In the current experiment, we estimate that with the present setup we could achieve a fractional advance of 0.55, with a

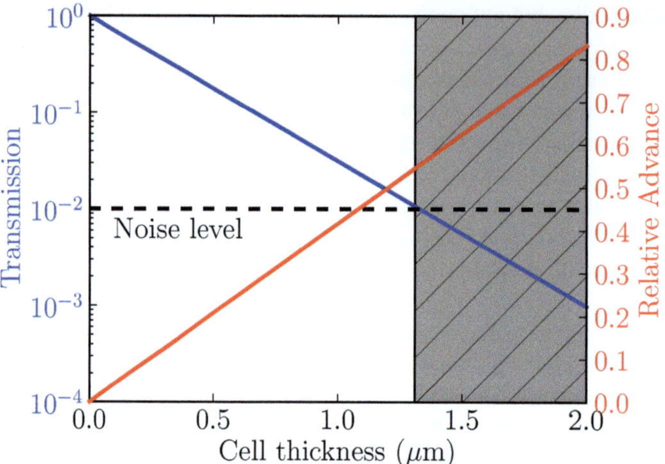

Fig. 7.7 Maximum fractional advance, based on current experimental signal-to-noise ratio. The transmitted intensity scales exponentially with the cell thickness (*blue*), while the fractional advance scales linearly. This limits the largest fractional advance that we could observe to around 0.55, which could be observed with a 1.3 μm thickness cell

transmission of 1 %, which could be implemented with a cell with a thickness $\ell \approx$ 1.3 μm. In principle the SNR can be increased by increasing the data acquisition time, but since the noise statistics are poissonian, the SNR scales as $\sqrt{N} \propto \sqrt{t_{acq}}$. The relative scaling of attenuation and fractional advance mean that returns are diminishing; a 100-fold increase in the acquisition time would mean a SNR $\sim$1,000, corresponding to a cell length of 2 μm for 0.1 % transmission, which yields a fractional advance of $\sim$0.85, an increase by a factor of only 1.5.

7.5 Outlook

Owing to the magnitude of the frequency-dependent refractive index in the medium, in the vicinity of the atomic resonances the group index of the medium varies over many orders of magnitude, giving rise to slow- and fast-light effects. On resonance we have demonstrated a significant fast-light effect, measuring a group index $n_g = (-1.0 \pm 0.1) \times 10^5$, the largest negative index measured to date. Due to the dipole–dipole interactions between atoms, which broaden the spectral lines, we can observe a significant fractional advance of sub-nanosecond optical pulses with very little distortion. The magnitude of the effect can be tuned by changing the atomic density and we can move from a fast-light to slow-light region by altering the frequency of the laser.

Whilst the high attenuation of the output pulses limits the potential applications of the current system, a relatively simple modification could prove fruitful. If the populations in the system were inverted, with the use of a fast pump pulse before investigation by the weak probe pulse, then the dynamics of the system change completely. Because the susceptibility depends on the degree of population inversion [Eqs. (2.16) and (2.21)], we could switch the system from a attenuating fast-light medium (on resonance) to a amplifying tunable slow-light medium, which could be useful for optical communications systems, for example. Work towards this goal is the focus of Chap. 9.

References

1. R.W. Boyd et al., Slow light and its applications. J. Opt. Soc. Am. B **25**, SL1 (2008)
2. T.F. Krauss, Why do we need slow light? Nat. Photon. **2**, 448 (2008)
3. M. Lukin, A. Imamoğlu, Nonlinear optics and quantum entanglement of ultraslow single photons. Phys. Rev. Lett. **84**, 1419 (2000)
4. Y. Wu, L. Deng, Ultraslow optical solitons in a cold four-state medium. Phys. Rev. Lett. **93**, 143904 (2004)
5. R.W. Boyd, Slow and fast light: fundamentals and applications. J. Mod. Opt. **56**, 1908 (2009)
6. R. de Kronig, On the theory of dispersion of X-rays. J. Opt. Soc. Am. **12**, 547 (1926)
7. H. Kramers, La diffusion de la lumiere par les atomes. In *Atti Cong. Intern. Fisica* (*Transactions of Volta Centenary Congress*) vol. 2, pp. 545, (1927)
8. C.A. Brau, *Modern Problems in Classical Electrodynamics* (Oxford University Press, Oxford, 2004)
9. M.L. Boas, *Mathematical Methods in the Physical Science*, 3rd edn. (Wiley, London, 2006)
10. G. Brooker, *Modern Classical Optics* (Oxford University Press, Oxford, 2003)
11. L.F. Mollenauer, R.H. Stolen, J.P. Gordon, Experimental observation of picosecond pulse narrowing and solitons in optical fibers. Phys. Rev. Lett. **45**, 1095 (1980)
12. R.J.P. Engelen et al., The effect of higher-order dispersion on slow light propagation in photonic crystal waveguides. Opt. Express **14**, 1658 (2006)
13. P. Siddons, N.C. Bell, Y. Cai, C.S. Adams, I.G. Hughes, A gigahertz-bandwidth atomic probe based on the slow-light Faraday effect. Nat. Photon. **3**, 225 (2009)
14. J. Ranka, R. Schirmer, A. Gaeta, Observation of pulse splitting in nonlinear dispersive media. Phys. Rev. Lett. **77**, 3783 (1996)
15. P. Siddons, C.S. Adams, I.G. Hughes, Off-resonance absorption and dispersion in vapours of hot alkali-metal atoms. J. Phys. B **42**, 175004 (2009)
16. R.M. Camacho, M.V. Pack, J.C. Howell, A. Schweinsberg, R.W. Boyd, Wide-bandwidth, tunable, multiple-pulse-width optical delays using slow light in cesium vapor. Phys. Rev. Lett. **98**, 153601 (2007)
17. S.E. Harris, J.E. Field, A. Imamoğlu, Nonlinear optical processes using electromagnetically induced transparency. Phys. Rev. Lett. **64**, 1107 (1990)
18. K.-J. Boller, A. Imamoğlu, S. Harris, Observation of electromagnetically induced transparency. Phys. Rev. Lett. **66**, 2593 (1991)
19. M.O. Scully, Enhancement of the index of refraction via quantum coherence. Phys. Rev. Lett. **67**, 1855 (1991)
20. M. Fleischhauer, A. Imamoğlu, J. Marangos, Electromagnetically induced transparency: optics in coherent media. Rev. Mod. Phys. **77**, 633 (2005)
21. J. Gea-Banacloche, Y.-Q. Li, S.-Z. Jin, M. Xiao, Electromagnetically induced transparency in ladder-type inhomogeneously broadened media: theory and experiment. Phys. Rev. A **51**, 576 (1995)
22. O. Schmidt, R. Wynands, Z. Hussein, D. Meschede, Steep dispersion and group velocity below c/3000 in coherent population trapping. Phys. Rev. A **53**, R27 (1996)

23. M. Bouchiat, J. Brossel, Relaxation of optically pumped Rb atoms on paraffin-coated walls. Phys. Rev. **147**, 41 (1966)
24. A. Sargsyan, D. Sarkisyan, U. Krohn, J. Keaveney, C.S. Adams, Effect of buffer gas on electromagnetically induced transparency in a ladder system using thermal rubidium vapor. Phys. Rev. A **82**, 045806 (2010)
25. M. Erhard, H. Helm, Buffer-gas effects on dark resonances: theory and experiment. Phys. Rev. A **63**, 043813 (2001)
26. M. Kash et al., Ultraslow group velocity and enhanced nonlinear optical effects in a coherently driven hot atomic gas. Phys. Rev. Lett. **82**, 5229 (1999)
27. D. Budker, D. Kimball, S. Rochester, V. Yashchuk, Nonlinear magneto-optics and reduced group velocity of light in atomic vapor with slow ground state relaxation. Phys. Rev. Lett. **83**, 1767 (1999)
28. L.V. Hau, S.E. Harris, Z. Dutton, C.H. Behroozi, Light speed reduction to 17 metres per second in an ultracold atomic gas. Nature **397**, 594 (1999)
29. M. Fleischhauer, M.D. Lukin, Dark-state polaritons in electromagnetically induced transparency. Phys. Rev. Lett. **84**, 5094 (2000)
30. M.D. Lukin, S.F. Yelin, M. Fleischhauer, Entanglement of atomic ensembles by trapping correlated photon states. Phys. Rev. Lett. **84**, 4232 (2000)
31. C. Liu, Z. Dutton, C.H. Behroozi, L.V. Hau, Observation of coherent optical information storage in an atomic medium using halted light pulses. Nature **409**, 490 (2001)
32. D. Phillips, A. Fleischhauer, A. Mair, R. Walsworth, M. Lukin, Storage of light in atomic vapor. Phys. Rev. Lett. **86**, 783 (2001)
33. A. Browaeys, Catch and release of photons. Physics **6**, 25 (2013)
34. Y.O. Dudin, L. Li, A. Kuzmich, Light storage on the time scale of a minute. Phys. Rev. A 031801(R) (2013)
35. D. Maxwell et al., Storage and control of optical photons using Rydberg polaritons. Phys. Rev. Lett. **110**, 103001 (2013)
36. A. Einstein, On the electrodynamics of moving bodies (translation). Ann. Phys. **7**, 891 (1905)
37. A. Sommerfeld, An objection against the theory of relativity of electrodynamics and its removal (translation). Phys. Z. **8**, 841 (1907)
38. A. Kuzmich, A. Dogariu, L.J. Wang, P.W. Milonni, R.Y. Chiao, Signal velocity, causality, and quantum noise in superluminal light pulse propagation. Phys. Rev. Lett. **86**, 3925 (2001)
39. M.D. Stenner, D.J. Gauthier, M.A. Neifeld, The speed of information in a 'fast-light' optical medium. Nature **425**, 695 (2003)
40. R.W. Boyd, Z. Shi, P.W. Milonni, Noise properties of propagation through slow- and fast-light media. J. Opt. **12**, 104007 (2010)
41. A.M. Akulshin, R.J. McLean, Fast light in atomic media. J. Opt. **12**, 104001 (2010)
42. C. Garrett, D. McCumber, Propagation of a Gaussian light pulse through an anomalous dispersion medium. Phys. Rev. A **1**, 305 (1970)
43. M. Crisp, Concept of group velocity in resonant pulse propagation. Phys. Rev. A **4**, 2104 (1971)
44. S. Chu, S. Wong, Linear pulse propagation in an absorbing medium. Phys. Rev. Lett. **48**, 738 (1982)
45. L.J. Wang, A. Kuzmich, A. Dogariu, Gain-assisted superluminal light propagation. Nature **406**, 277 (2000)
46. A. Akulshin, A. Cimmino, A. Sidorov, P. Hannaford, G. Opat, Light propagation in an atomic medium with steep and sign-reversible dispersion. Phys. Rev. A **67**, 2 (2003)
47. G.M. Gehring, A. Schweinsberg, C. Barsi, N. Kostinski, R.W. Boyd, Observation of backward pulse propagation through a medium with a negative group velocity. Science **312**, 895 (2006)
48. R.W. Boyd, P. Narum, Slow- and fast-light: fundamental limitations. J. Mod. Opt. **54**, 2403 (2007)
49. B. Macke, B. Ségard, F. Wielonsky, Optimal superluminal systems. Phys. Rev. E **72**, 035601 (2005)
50. H. Tanaka et al., Propagation of optical pulses in a resonantly absorbing medium: observation of negative velocity in Rb vapor. Phys. Rev. A **68**, 053801 (2003)

Chapter 8
Fluorescence Lifetime

In Chap. 5, we measured the cooperative Lamb shift in a thin Rb layer. Along with the shift, we observed a sharp transition in the line broadening coefficient, the mechanism of which is not well understood. However, theory suggests that given the presence of a cooperative shift in the system, there should also be a corresponding change in the lifetime, known as superradiance [1]. Instead of measuring the frequency spectrum, we can investigate the change in lifetime in the time domain, using the fluorescence as a direct probe.

In this chapter we investigate the fluorescence from a mm-thickness vapour layer, and compare this to the fluorescence signal in the nano-cell. At high atomic density, radiation trapping becomes important in the mm-thickness regime, whereas for the nano-cell the decay is dominated by wall-to-wall collisions.

8.1 Experimental Setup

The experimental setup is similar to that used in the previous chapter; a Pockels cell generates nanosecond pulses from a CW laser source. The excitation laser is stabilised using polarisation spectroscopy to an atomic resonance—either the $F_g = 1 \rightarrow F_e = 2$ or $F_g = 2 \rightarrow F_e = 3$ transition in ^{87}Rb. The pulse intensity is strong enough to significantly excite the medium, with a peak optical power of around $100\,\mu$W focussed to a spot size of $\sim$30 μm, and the off-axis fluorescence is observed with a single photon counting module (SPCM). A histogram of the photon arrival times is then recorded. The data contains the signal with count rate C_s, the background light in the room (minimised as much as possible but still the main source of noise) C_{bg}, and the dark count rate of the detector C_{dc}, which is the amount of counts when there is no light present. This has been measured for this detector to be around 50 Hz, uniformly distributed in time (see Appendix C). The room background rate varies over different experiments, but remains constant over the timescale of a single recording, and is of the order of a few hundred Hz, significantly less than the signal count rate, which is typically in the tens of kHz (CW equivalent).

J. Keaveney, *Collective Atom–Light Interactions in Dense Atomic Vapours*, Springer Theses, DOI: 10.1007/978-3-319-07100-8_8,
© Springer International Publishing Switzerland 2014

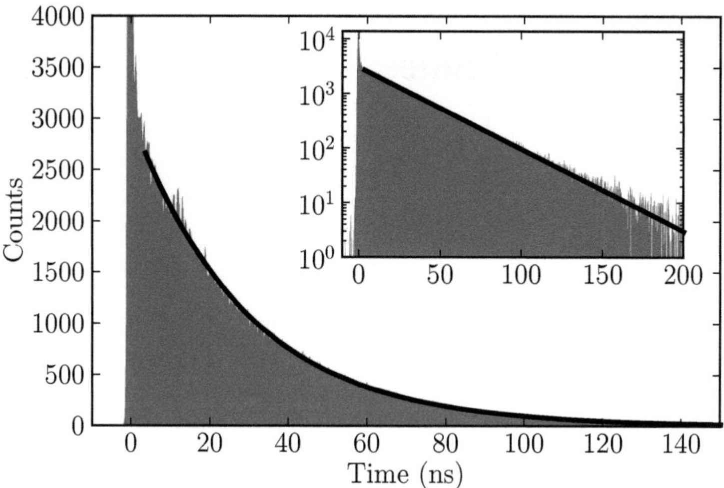

Fig. 8.1 Fluorescence decay of the 5P$_{3/2}$ state, measured in a 4 mm-thickness cell at 28 °C. The counts are per bin, and the bin width of the histogram is 250 ps. A nanosecond pulse is incident on the medium at $t \approx 0$, and we observe a large initial count rate, due to scatter from the cell heater and windows. After the initial spike, a least-squares fit to a decaying exponential yields a lifetime $\tau = 26.9 \pm 0.5$ ns

8.2 Fluorescence Lifetime in the 4 mm Cell

As a proof-of-principle test, an experiment was set up using a 4 mm thickness vapour cell to measure the well-known excited state lifetime of the 5P$_{3/2}$ state of Rb, before moving on to the nano-cell.

In order to determine the natural lifetime of the excited state, a relatively cold vapour is needed to avoid significant atomic interactions, as discussed in Chap. 5, or radiation trapping caused by large optical depths, which will be discussed later in this chapter.

Figure 8.1 shows the measured fluorescence signal as a function of time after excitation. The inset to the plot shows the same data on a log scale. The sharp peak at $t = 0$ is the light from the pulse that is scattered from the glass of the cell and the cell heater (we still observe this spike when the pulse is incident on a region of the cell with no Rb present). The constant noise floor due to background light and dark counts is taken as the mean number of counts in the time before the pulse arrives ($t < -3$ ns), has been subtracted from the data.

The data are fit to an exponential decay $C(t) = C_0 e^{-t/\tau}$, where $C(t)$ is the number of counts at time t and $C_0 = C(t = 0)$. For the data presented in Fig. 8.1 the lifetime τ directly corresponds to the excited state lifetime, with a fitted value of $\tau = 26.9 \pm 0.5$ ns, in reasonable agreement with accepted value of 26.2 ns [2]. The slightly larger mean value is probably a result of a small but finite amount of radiation trapping, similar to results in Ref. [3]. When plotted on a semilog scale, the data fit well to a straight line. As a more quantitative indicator, the reduced χ^2 value for the fit is 1.6, indicating a good fit [4]. The black line in the figure is the fit to the data. We fit only the data from $t \geq 5$ ns, to ignore the scattered light around $t = 0$, and assume Poissonian statistics in the number of counts.

8.2.1 Multiple-Scattering Events

If the optical depth is large enough, an emitted photon can be reabsorbed before it leaves the medium. These multiple scattering events are ubiquitous in nature; for example, in a cloud, the density of the water vapour changes its opacity because the light is scattered many times before it leaves the cloud [5]. This behaviour is usually diffusive, with the associated Brownian motion having a well defined mean free path $\langle x \rangle$ and variance $\langle x^2 \rangle$ determined by Gaussian statistics (the central limit theorem [4]).

However, for a number of systems the behaviour is super-diffusive, dominated by the rare but long path steps in the tail of the distribution known as Lévy flights. These long- (or heavy-) tailed distributions are characterised by a step-size probability that follows a power law, $P(x) \propto 1/x^{\alpha}$. When $\alpha < 3$, the variance of the distribution diverges, and hence the statistics no longer necessarily follow the central limit theorem. Many systems from a wide range of disciplines have been shown to display these long-tailed distributions, including in ecology, physiology, financial systems and the physical sciences (see [6] and references therein). The multiple scattering of resonant light from hot atoms is a natural (i.e. not engineered) example of this phenomenon.

Physically, the origin of the long-tailed distribution is in the inhomogeneities of the system. In many cases, these are spatial inhomogeneities, for example the variation in the size of scatterers in non-resonant systems [7]. In resonant fluorescence however, it is the spectral inhomogeneity that is important. In particular, it is the frequency redistribution of the incident (quasi-monochromatic) light by the atoms, which have some spectral width, which causes an effective inhomogeneity in the optical depth of the (quasi-homogeneous) medium-light that is re-emitted in the wings of the distribution will have a much lower probability to be reabsorbed and hence travels much further than light that is emitted near resonance. The difference in scattering between monochromatic light (exponential decay following Beer's law) as compared to light already scattered by atoms (and hence frequency redistributed) was demonstrated by Mercadier et al. [8]. By measuring the fluorescence from this pre-scattered light they were able to infer the step-size distribution, confirming the probability distribution with $\alpha = 2.41 \pm 0.12$, characteristic of Lévy flights.

The multiple scattering events all increase the measured lifetime of the light in the medium; the light is briefly 'trapped' or 'imprisoned' in the medium before it escapes. Much of the theoretical work in this field was derived by Holstein [9, 10] based on a sum of decaying exponentials (now called Holstein modes). However, Holstein's results assumed high density media and fail to account for the low density

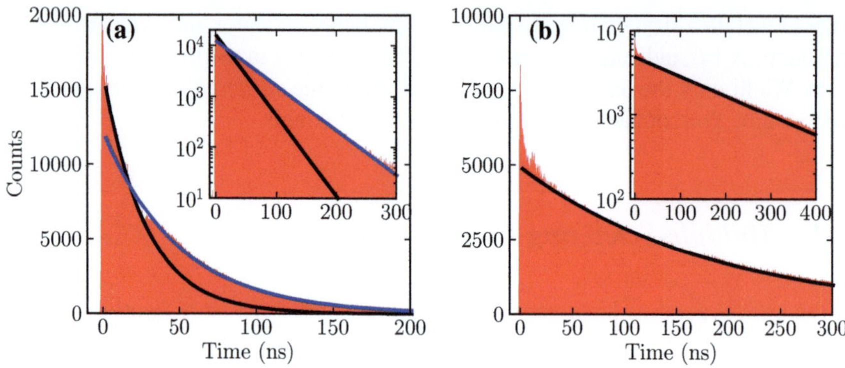

Fig. 8.2 Fluorescence decay from the 4 mm cell, showing radiation trapping as density is increased. **a** At 75 °C, the data fits a double exponential decay, the first (*black*) with the excited state lifetime and then a second longer decay (*blue*). **b** As density is increased further ($T = 110$ °C), only the longer decay remains. The counts are per bin, and the bin width of the histogram is 250 ps

behaviour which the current work is concerned with. At low densities, assuming a simple two-level model with a single excited state that in the absence of any radiation trapping has a lifetime τ_0, the effect of radiation trapping is to increase the lifetime linearly with density [11, 12]

$$\tau = \tau_0(1 + aB\,\alpha L), \tag{8.1}$$

where a is a geometry-dependent scaling factor which we take as 1 for a cylinder [13], for a closed system the branching ratio $B = 1$, and αL is the effective optical depth of the medium which differs from the normal optical depth because the detection is off-axis. L is effectively the distance from the point of excitation to the edge of the medium in the direction of the detector, which we estimate for our geometry to be ~5 mm. For resonant excitation, α is the absorption coefficient of the zero-velocity class and can be expressed as [13]

$$\alpha = \frac{N\lambda^3}{8\pi}\frac{\Gamma_0}{u\pi^{1/2}}, \tag{8.2}$$

where u is the most probable thermal velocity and $\Gamma_0 = 1/\tau_0$ is the excited state decay rate.

Clearly, for the high density vapour in the nano-cell radiation trapping could be a significant factor. In order to understand this process, we first investigate the time-dependent fluorescence in the 4 mm cell as we raise the optical depth. Figure 8.2 shows the effect of radiation trapping for two temperatures. Panel (a) shows data at $T = 75$ °C ($N \approx 10^{12}$ cm^{-3}), whilst panel (b) shows data at $T = 110$ °C ($N \approx 10^{13}$ cm^{-3}). For the lower density, we observe two distinct decay features, the first of which corresponds to the excited state lifetime. The second we attribute to radiation trapping, and has a lifetime $\tau = 47.7 \pm 0.1$ ns. At the higher density we see only the longer radiation trapped lifetime, in this case with a lifetime $\tau = 129.6 \pm 0.4$ ns. At

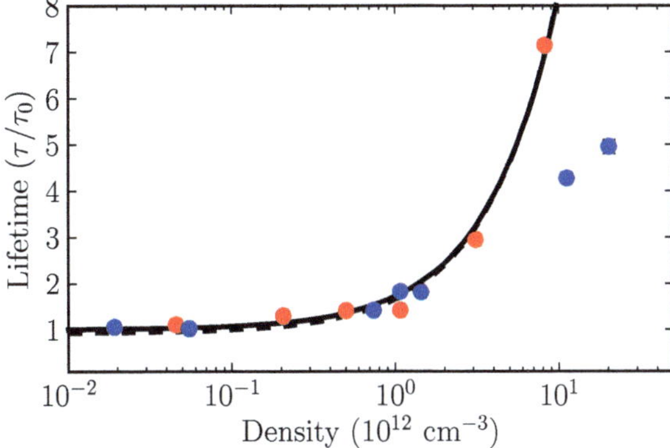

Fig. 8.3 Fluorescence lifetime, in units of the natural lifetime τ_0, plotted against density. *Blue* points are data with the pulse carrier detuning on resonance with the ^{87}Rb $F_g = 2 \rightarrow 3$ transition on the D2 line, while *red* points are for the laser resonant with the ^{87}Rb $F_g = 1 \rightarrow 2$ transition. The *solid line* is a linear fit to the data (excluding the two *largest blue* data points), with an offset corresponding to a measurement of the natural lifetime of the excited state of $\tau_0 = 24 \pm 2$ ns. Error bars calculated from the fits are smaller than the data point markers

the lower density, the light that reaches the detector soonest is that which undergoes no additional scattering.

For the relatively low densities used here, the effects of dipole-dipole interactions are small ($V_{dd} < \Gamma$, see Chap. 5) and therefore to a good approximation the optical depth is proportional to the density. We characterise the radiation trapping behaviour as a function of the density of the sample, and plot the results in Fig. 8.3. We do this for two cases where the pulse carrier frequency is resonant with the two ground states, as discussed earlier in the chapter. Excluding the two highest density points for the (open) $F_g = 2$ transition, the data fit to a linear function. The deviation of the two highest density points is probably a result of decay into the lower hyperfine ground state. By allowing the offset to vary we can infer the zero-density limit value for τ_0, which we find is 24 ± 2 ns, in reasonable agreement with our previous measurement and the literature value. In this case we obtain a fitted gradient of $1.89 \pm 0.04 \times 10^{-11}$ ns cm^3 (dashed line on the figure). Alternately if τ_0 is fixed at a value of 26.2 ns, we find a similar gradient from the fit of $1.93 \pm 0.05 \times 10^{-11}$ ns cm^3 (solid line on the figure). Using Eqs. 8.1 and 8.2 with $a = 1$, $B = 1$ and $L = 5$ mm, we obtain a coefficient of $\sim 8 \times 10^{-12}$ ns cm^3, close to the experimental value. The bandwidth of the pulse means that we excite more than just the zero velocity class atoms, and this could account for the difference.

8.3 Fluorescence Lifetime in the Nano-Cell

In the nano-cell, one may think that because the vapour is so thin in the propagation direction, radiation trapping effects may be negligible. In this case, the excitation would need to be off-axis, with the detector placed normal to the windows. However,

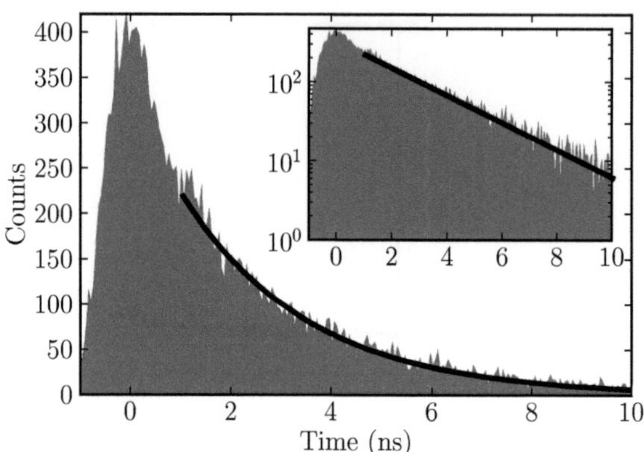

Fig. 8.4 Histogram of fluorescence counts in the nano-cell, with a thickness $\ell = 3\lambda/2$ and temperature $T = 160\,°C$. The bin width is 50 ps. The measured fluorescence fits to an exponential decay with lifetime $\tau = 2.6 \pm 0.2$ ns

with the current heater design, optical access is extremely limited and the only way of exciting off-axis is from the side, at 90° to the propagation direction of the laser. With the excitation from the side, we excite a wide range of cell thicknesses which is undesirable; a solution to this is to image only a specific region, using the imaging system to determine the local thickness. In this work however, for simplicity we excite in the same way as for the 4 mm cell (at normal incidence to the windows), and detect from the side. As we shall show, radiation trapping is not the dominant effect in this system.

Whilst the cooperative Lamb shift (CLS) only becomes apparent when the density is very high, the change in the broadening coefficient appears to happen for all densities. One may expect, in absence of any radiation trapping, a lifetime modification due to the dipole-dipole interaction to

$$\frac{1}{\tau} = \frac{1}{\tau_0} + \Gamma_{dd}. \tag{8.3}$$

At high densities where the CLS becomes important we therefore expect a lifetime that is much shorter than the pulse, since Γ_{dd} is of the order of GHz. However, the enhancement in the broadening coefficient that we also observe as a function of cell thickness is present from very low densities, therefore it is interesting to look at the fluorescence lifetime at relatively low densities, where we still expect a reasonably long lifetime—shorter than the natural lifetime but longer than the pulse width.

Figure 8.4 shows the measured fluorescence signal from the nanocell, with a cell thickness $\ell = 3\lambda/2$. At a temperature $T = 160\,°C$, we have a number density $N = 1.6 \times 10^{14}$ cm^{-3}, which gives $\Gamma_{dd} = 2\pi \times 16$ MHz, yielding an expected interaction-shortened lifetime of $\tau \approx 7$ ns. The atom-surface interaction for this cell thickness is negligible. However, we observe a decay that is commensurate with a lifetime of $\tau = 2.6 \pm 0.2$ ns. In order for this to be due to dipole-dipole interactions, we would require a temperature $T = 190\,°C$ ($N = 6.2 \times 10^{14}$ cm^{-3}), too high to be due to inaccuracy in the temperature measurement. As confirmation of this, we

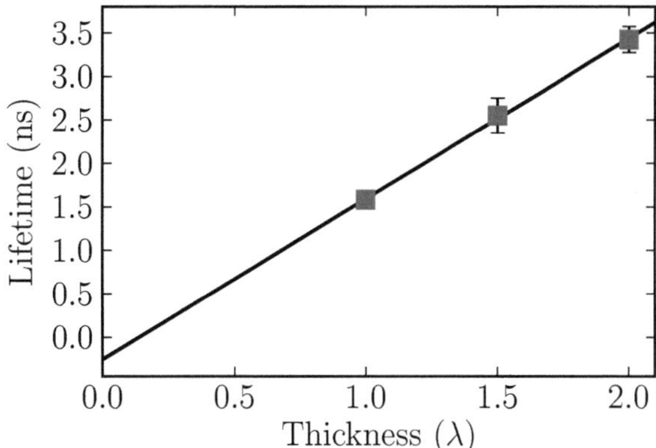

Fig. 8.5 Fluorescence lifetime versus cell thickness in the nano-cell. We find a linear relationship between the lifetime and the cell thickness, implying that the lifetime is limited by the time of flight of the atoms across the cell

measure the same lifetime, within error, at a temperature of $T \approx 130\,^\circ$C. Exciting the medium from the side with detection normal to the windows also yields a similarly short lifetime.

We attribute the lifetime to the atomic time-of-flight across the cell. Unlike with CW spectroscopy, where the atoms moving parallel to the windows interact for longer than the atoms moving wall-to-wall, the short excitation pulse means that all atoms interact with the beam for the same amount of time. This makes the experiment much more sensitive to the wall-to-wall atoms than is the case with CW excitation, as there are many more atoms that move wall-to-wall than parallel to the windows.

To confirm this mechanism, we look at the measured fluorescence lifetime as a function of the vapour thickness. As expected, the lifetime increases linearly with thickness. For thermal atoms, we expect the atoms to be moving with most probable velocity $u \approx 300$ m/s, which would correspond to a lifetime $\tau = \ell/u$. The measured gradient is consistent with $\tau \sim \ell/1.3u$, and the slope has a negative offset. The negative offset can be understood as an effective shortening of the cell, due to strong atom-surface interactions in the vicinity of the windows—the atom decays before it hits the wall. Extrapolating from the fit, we find the offset is consistent with the vapour being effectively shortened by $\sim$100 nm or 50 nm on either window which is a reasonable estimate of the distance at which atom-surface interactions become important (see Chap. 4).

8.4 Outlook

As measuring the time-domain fluorescence is a simple and relatively inexpensive method for studying the radiative lifetime, it could be possible to perform a precision measurement of the excited state lifetime, if the experimental parameters, particularly the temperature of the cell, could be controlled with greater precision, as in principle the signal to noise ratio is limited only by the Poissonian counting noise which increases with the square root of the accumulation time (for a constant signal rate).

The original aim of this experiment was to observe directly the enhancement of the decay rate that is the 'smoking gun' of a superradiant process. However, the extremely short lifetime of the wall-to-wall atoms is the dominant feature at low density. In order to circumvent this issue, we could move to higher density where the interactions are on a shorter timescale than wall-to-wall time-of-flight, but in this case, the decay time including collisions is expected to be shorter than the pulse width. A shorter optical pulse would partially solve this problem—moving to a 100 ps pulse may be an idea for future experiments, but this comes at the cost of higher pulse bandwidth, which would be comparable to the width of the entire absorption line, and also increases the expense since a better detection system would be needed, in addition to a different laser source.

Using a multilevel excitation scheme to Rydberg states, the same technique has been applied to successfully observe superradiant cascade using thermal Cs vapour [14], but it seems very unlikely that a direct observation of a superradiant decay is possible in the current system.

Despite these setbacks, we still hope that this time-domain fluorescence technique can probe some interesting physics in the nanocell. One could also imagine using a strong pulse to drive Rabi oscillations in the medium, observing the variation in the decay as either the power or the duration of the pulse is varied, similar to Ref. [15].

In the mm-cell, the radiation trapping process is prominent at moderate densities. The double-exponential decay of the fluorescence could be a sign of Lévy flights in the decay and it would be interesting to study this behaviour further.

References

1. R. Dicke, Coherence in spontaneous radiation processes. Phys. Rev. **93**, 99 (1954)
2. D.A. Steck, Rubidium 85 D Line Data (2009)
3. N.D. Zameroski, W. Rudolph, G.D. Hager, D.A. Hostutler, A study of collisional quenching and radiation-trapping kinetics for Rb(5p) in the presence of methane and ethane using time-resolved fluorescence. J. Phys. B **42**, 245401 (2009)
4. I.G. Hughes, T.P.A. Hase, *Measurements and their Uncertainties: A Practical Guide to Modern Error Analysis* (OUP, Oxford, 2010)
5. L.R. Bissonnette et al., LIDAR multiple scattering from clouds. Appl. Phys. B **60**, 355 (1995)
6. M. Chevrollier, Radiation trapping and Lévy flights in atomic vapours: an introductory review. Contemp. Phys. **53**, 227 (2012)
7. P. Barthelemy, J. Bertolotti, D.S. Wiersma, A Lévy flight for light. Nature **453**, 495 (2008)
8. N. Mercadier, W. Guerin, M. Chevrollier, R. Kaiser, Lévy flights of photons in hot atomic vapours. Nature Phys. **5**, 602 (2009)
9. T. Holstein, Imprisonment of resonance radiation in qases. Phys. Rev. **72**, 1212 (1947)
10. T. Holstein, Imprisonment of resonance radiation in gases. II. Phys. Rev. **83**, 1159 (1951)
11. H. Holt, Resonance trapping at low pressures. Phys. Rev. A **13**, 1442 (1976)
12. J. Klose, Mean life of the $4s^2 S_{1/2}$ resonance level in Al I. Phys. Rev. A **19**, 678 (1979)
13. J. Klose, Mean life of the 27,887-cm^{-1} level in U$_i$. Phys. Rev. A **11**, 1840 (1975)
14. C. Carr, R. Ritter, K.J. Weatherill, C.S. Adams, Cooperative non-equilibrium phase transition in a dilute thermal atomic gas. arxiv.org, 1302.6621 (2013). arXiv:1302.6621v1
15. B. Darquié et al., Controlled single-photon emission from a single trapped two-level atom. Science **309**, 454 (2005)

Chapter 9
Coherent Dynamics

In Chap. 7 we showed that in the weak probe limit, where we assume all of the population remains in the ground state, the group index of a dense Rb atomic vapour can be very large. On resonance, we observe a group index of -10^5, which varies by a relatively small amount over a GHz-bandwidth region. This allowed for superluminal propagation of sub-nanosecond optical pulses with very little distortion.

In this chapter instead of only weakly probing the ground-state medium, we drive significant population into the excited state with a second pump laser. By doing this we change the propagation dynamics of the weak probe pulse, since the susceptibility that is probed depends on the amount of excitation in the medium. If all of the population is driven into the excited state and probed before the medium decays (either through spontaneous emission or interactions with other atoms or the cell windows), then in principle the medium that the probe experiences can be changed from a fast-light to a slow-light medium, and instead of attenuation, the population inversion amplifies the probe pulse as it propagates.

The ability to do this crucially depends on the coherent evolution of the system, and in this chapter we present initial results from an ongoing experiment where we probe sub-nanosecond timescale coherent dynamics in the nanocell.

9.1 Introduction

In a two-level atomic system, as discussed in Chap. 2, if the driving Rabi frequency Ω is large compared to the decay rate of the excited state Γ then the populations of the ground and excited states can oscillate many times before reaching a steady-state. We plot in Fig. 9.1 the evolution of the excited state population ρ_{22} as a function of time for various values of the ratio Ω/Γ. When $\Omega/\Gamma \ll 1$, there is little oscillation and the system remains mostly in the ground state. However, as the ratio increases, the oscillations become significant, and a transient population inversion can be realised. When this happens, the coherence between the levels represented by the off-diagonal

J. Keaveney, *Collective Atom–Light Interactions in Dense Atomic Vapours*, Springer Theses, DOI: 10.1007/978-3-319-07100-8_9,

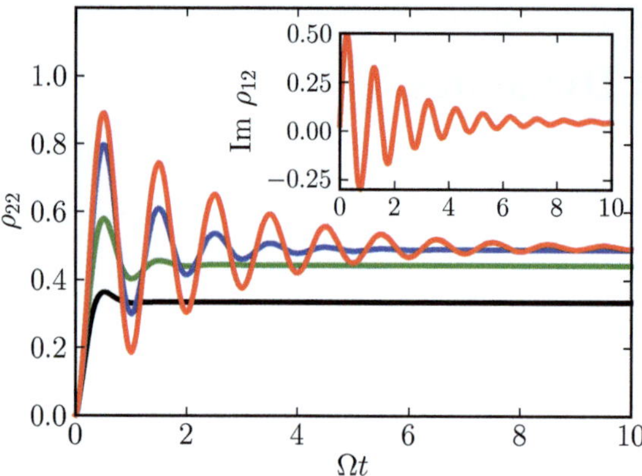

Fig. 9.1 Rabi oscillations plotted for various ratios of the Rabi frequency to damping rate; $\Omega/\Gamma = 1$ (*black*), 2 (*green*), 5 (*blue*) and 10 (*red*). The inset shows the oscillation in the imaginary part of the coherence, which is responsible for absorption. When $\mathrm{Im}(\rho_{12})$ is negative, we expect amplification instead of absorption

elements of the density matrix ρ_{12} switches sign. If we recall from Eq. 2.21 the fundamental link between the coherence term and the atomic susceptibility, we see immediately that the population inversion dramatically alters the optical properties of the atom. By switching the sign of ρ_{12} we switch from an absorbing medium into an amplifying medium; physically this comes about because the laser now drives more population down from the excited state, with emission of a photon, than up from the ground state with the absorption of a photon.

In an ideal experiment, the medium would be pumped with a strong optical pulse, before probing with a second weak pulse some time later. Whilst in principle the pump and probe can be the same wavelength, in practice this is not feasible in our experiment, since there are many orders of magnitude in power between them, separating the probe signal from the pump light is challenging. If the two pulses are well separated this becomes easier, but in the present experiment the effective decay time of the excited state is very fast, on the order of a few nanoseconds, due to collisions between the thermal atoms and the windows of the cell (see Chap. 7). The decay rate in this high density regime is expected to become very large, so equivalently large Rabi frequencies are needed to observe any coherent behaviour, on the order of 1 GHz or more.

Given the combined need for a multi-level excitation scheme to separate pump and probe pulses, and for large Rabi frequencies, a Vee-type excitation scheme was chosen. The pump pulse couples a ground state to the $5P_{3/2}$ manifold, whilst the probe is resonant with a transition from the same ground state to an energy level in the $5P_{1/2}$ manifold. Similar multilevel systems have previously demonstrated GHz Rabi oscillations in thermal Rb cells with a ladder scheme, with the pump tuned to a transition from the $5P_{3/2}$ level up to a Rydberg state [1, 10]. However, the optical power required to obtain GHz Rabi frequencies for this transition is very large. Using

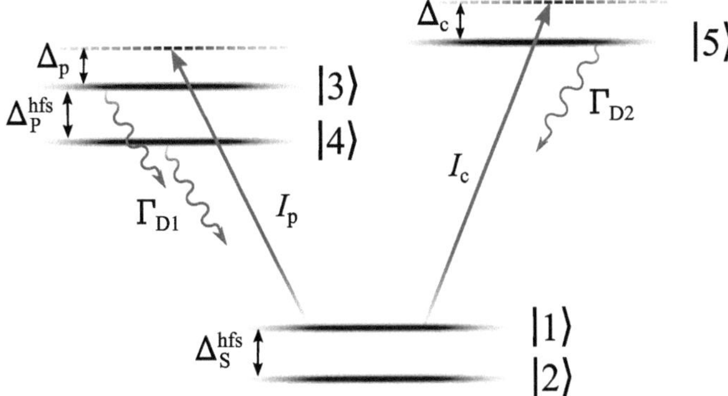

Fig. 9.2 Energy level scheme of the five level system. The states $|1\rangle$ to $|5\rangle$ correspond to, in order, the $F = 3$ and $F = 2$ states of $5S_{1/2}$, the $F = 3$ and $F = 2$ levels of $5P_{1/2}$, and the $5P_{3/2}$ manifold of ^{85}Rb. The states are coupled with two lasers: the probe laser with intensity I_p couples the $5S_{1/2}$ and $5P_{1/2}$ manifolds, whilst the coupling laser with intensity I_c couples $5S_{1/2}$ to $5P_{3/2}$. Other symbols are defined in the main text

another ground state transition for the pump is advantageous because the transition is very strong, allowing large Rabi frequencies for moderate amounts of laser power. In our experiment we can achieve GHz Rabi frequencies with around 20 mW of optical power.

9.2 Optical Bloch Simulations

The energy levels of the model system are shown in Fig. 9.2. The system forms a vee-type system of three manifolds—$5S_{1/2}$ (states $|1\rangle$ and $|2\rangle$), $5P_{1/2}$ (states $|3\rangle$ and $|4\rangle$), which decays naturally at a rate Γ_{D1}, and $5P_{3/2}$ (state $|5\rangle$), which decays naturally at a rate Γ_{D2}. The D1 laser with a wavelength 795 nm is the CW probe, whilst the D2 laser at 780 nm is the pump laser, and can be either CW or pulsed. Within the manifolds we consider the hyperfine splitting of both the ground and excited states, since the Rabi frequencies we consider are of the same order as the ground state hyperfine splitting, and much larger than the excited state splitting. However, as the pulse bandwidth is larger than the $5P_{3/2}$ hyperfine splitting, we consider this as one state.

The Hamiltonian of this system is given by

$$
\mathcal{H}\partial = \frac{\hbar}{2}
\begin{pmatrix}
0 & 0 & \Omega_{13} & \Omega_{14} & \Omega_{15} \\
0 & 2\Delta_S^{\mathrm{hfs}} & \Omega_{23} & \Omega_{24} & \Omega_{25} \\
\Omega_{13} & \Omega_{23} & -2\Delta_p & 0 & 0 \\
\Omega_{14} & \Omega_{24} & 0 & -2(\Delta_p - \Delta_P^{\mathrm{hfs}}) & 0 \\
\Omega_{15} & \Omega_{25} & 0 & 0 & -2\Delta_c
\end{pmatrix},
\tag{9.1}
$$

where the levels i and j are coupled with Rabi frequencies Ω_{ij}. The strength of the coupling depends on the intensities of the two lasers and the transition strengths for the indivdual hyperfine transitions, which can be found tabulated in Refs. [2, 3]. In practice, we specify only one of each of the probe and coupling Rabi frequencies, and the others are derived from these. The decay matrix $\mathcal{L}$ contains the branching ratios for decay out of the excited states back to either of the ground states. We include the effect of self-broadening as a modification to the decay rate, by replacing $\Gamma_{\mathrm{D1,D2}}$ with $\Gamma_{\mathrm{D1,D2}} + \beta_{\mathrm{D1,D2}} N$, where $\beta_{\mathrm{D1, D2}}$ is the state-dependent self-broadening coefficient (see Chap. 5). We can include any additional dephasing into this term, if required. Finally, we include a dephasing term between the two ground states, γ_{21}, which depends on the vapour thickness, to simulate the collisions with the walls which repopulate the ground states evenly.

As the experimental pulse carrier frequency is resonant with the closed transition $F_g = 3 \rightarrow F_e = 4$ in $^{85}\mathrm{Rb}$, we assume that the transition $|1\rangle \rightarrow |5\rangle$ is closed, i.e. that state $|5\rangle$ does not decay into state $|2\rangle$. The simulation is relatively insensitive to changes in this value.

We can then use the Lindblad equation (Eq. 2.13) to find the time evolution of the density matrix, either in the steady state with a CW pump laser or in the transient regime with a pulsed pump laser.

9.2.1 Steady-State Solutions

With a CW pump laser, we find the steady-state solution in the same way as in Chap. 2. We include Doppler broadening by making the substitutions $\Delta_{\mathrm{p}} \rightarrow \Delta_{\mathrm{p}} - k_{\mathrm{p}} v$ and $\Delta_{\mathrm{c}} \rightarrow \Delta_{\mathrm{c}} - k_{\mathrm{c}} v$, where $k_{\mathrm{p,c}}$ are the wavevectors of the probe and coupling lasers, and integrating over the velocity distribution. The two lasers are copropagating, so the atom sees a Doppler shift from both that is in the same direction. This maximises the amount of atoms that encounter a two-photon resonance condition.

The transmission of the probe laser is proportional to the sum of all the probe transitions. In a similar manner to Eq. (2.21) we have

$$\mathrm{Im}(\chi) = \frac{2N\mu^2}{\hbar\epsilon_0} \sum_{i=1,2} \sum_{j=3,4} \frac{\mathrm{Im}(\rho_{ij})}{\Omega_{ij}}. \tag{9.2}$$

The transmission is found from $\mathcal{T} = \exp(-k\ell\,\mathrm{Im}(\chi))$ in the normal way.

With two CW lasers, it is often insightful to consider the dressed state picture, where the coupling laser splits the probe transition into two, which are separated by the coupling laser Rabi frequency, Ω_{c}, and centred on the energy level of the undressed or 'bare' state $|2\rangle$. Electromagnetically induced transparency (EIT) can then be understood as interference between the dressed states $|+\rangle$ and $|-\rangle$. The frequency splitting between these dressed states is equal to the coupling Rabi frequency. This splitting is known as Autler-Townes (AT) splitting. For large coupling

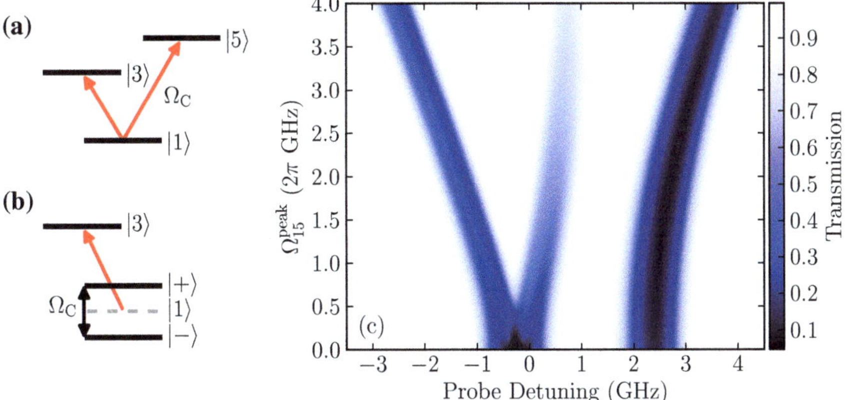

Fig. 9.3 Energy levels in the bare (**a**) and dresed (**b**) states, with a resonant coupling laser. The state common to both lasers, in this case the ground state, is 'dressed' by the coupling laser creating the new states $|+\rangle$ and $|-\rangle$. These two states are split by an amount equal to the Rabi frequency of the coupling laser. (**c**) Simulation results of the 5-level system, showing Autler-Townes splitting of the resonant ground state as the probe detuning is varied. The splitting increases linearly with the coupling Rabi frequency, until it becomes comparable to the ground state hyperfine splitting, where the other ground state also starts to shift

Rabi frequencies, the splitting is so large that the absorption curve is well described by 2 separate absorption lines, centred on $\Delta_p = \pm\Omega_c/2$, as demonstrated in Fig. 9.3b. Panel (c) of Fig. 9.3 shows the Doppler-broadened transmission spectrum of the 5-level system with CW lasers as a function of probe detuning and coupling Rabi frequency, with the coupling laser resonant with the $|1\rangle \rightarrow |5\rangle$ transition. AT splitting of the $|1\rangle \rightarrow |3,4\rangle$ transition is evident as the coupling Rabi frequency increases, which splits linearly until $\Omega_c \sim 2$ GHz then becomes nonlinear as the second hyperfine ground state $|2\rangle$ becomes involved in the process. This occurs because the coupling laser also dresses the $|2\rangle \rightarrow |5\rangle$ transition, but with a large detuning. As Ω_c increases, the $|2\rangle \rightarrow |3,4\rangle$ transition splits asymmetrically, eventually interfering with the resonant state. Hence the experimental CW spectrum showing AT splitting yields information about the coupling Rabi frequency of the system.

9.2.2 Pulsed Excitation

An optical pulse has a time-dependent intensity $I(t)$, giving rise to time dependent Rabi frequencies for the D2 transitions. To simulate this system, we split the problem into small time steps Δt and calculate the density matrix. The initial calculation assumes that population is evenly distributed within the ground state manifold, and each subsequent iteration uses the solution of the previous iteration. We use a Gaussian for the temporal shape of the pulse intensity whose FWHM matches the experimental

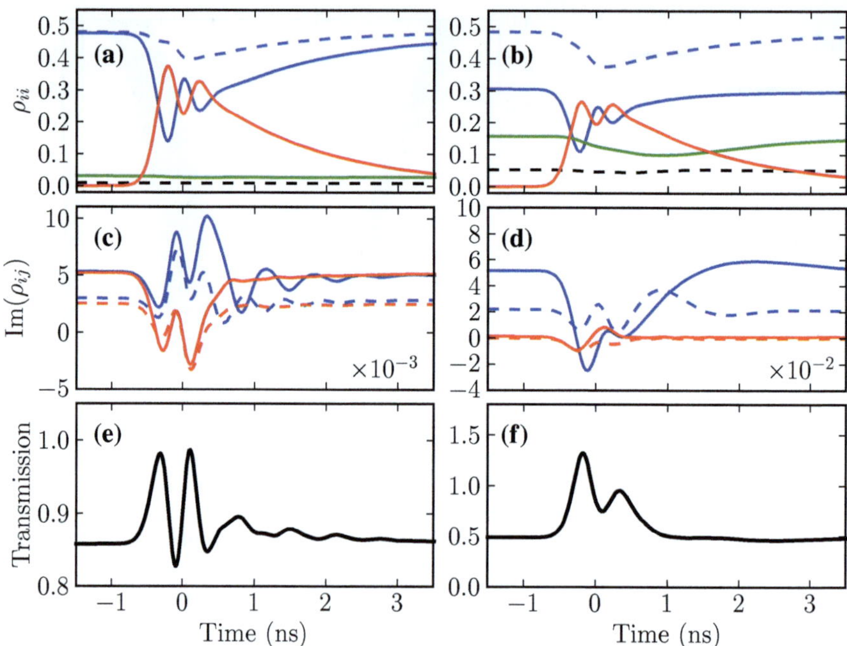

Fig. 9.4 Time evolution of the density matrix as calculated from the five-level optical Bloch model, with parameters $\Delta_p = 2\pi \times 1.5$ GHz (a, c, e), $\Delta_p = 0$ (b, d, f); $\Omega_{15}^{\text{peak}} = 2\pi \times 2.5$ GHz; $\Omega_{13} = 2\pi \times 500$ MHz; $\Delta_c = 0$; $\ell = 1\ \mu$m; T = 200 °C. Panels **a** and **b** show populations of the states: ρ_{11}, *solid blue*; ρ_{22}, *dashed blue*; ρ_{33}, *green*; ρ_{44}, *dashed black*; ρ_{55}, *red*. Panels **c** and **d** show the imaginary part of the probe coherences: ρ_{13}, *solid blue*; ρ_{14}, *dashed blue*; ρ_{23}, *solid red*; ρ_{24}, *dashed red*. Finally, panels **e** and **f** show transmission of the probe beam, calculated from Eq. (9.2) using the sum of the probe coherences

value. The pulse bandwidth is included as a dephasing term, similar to Refs. [4, 5]. As before, Doppler broadening can be included but it is time consuming and does not qualitatively alter the results of the simulation, so it is not included in the results we present below.

To begin, it is useful to look at the elements of the density matrix. Figure 9.4 shows two cases where the probe is resonant with the $|1\rangle \rightarrow |3\rangle$ transition (panels a, c, e) and where the probe is detuned midway between the two hyperfine ground states, $\Delta_p = 2\pi \times 1.5$ GHz. The probe laser is strong, with Rabi frequency $\Omega_{13} = 2\pi \times 500$ MHz. We calculate with a strong probe because the experiment also uses a strong probe, which is a consequence of using fast detectors. Their sensitivity is much lower than an equivalent low frequency photodiode, requiring the use of higher optical power. However, the simulations are relatively insensitive to the probe intensities. Qualitatively, the coherence terms do not vary significantly.

We plot the temporal evolution of (a, b) the populations and (c, d) the probe coherences, along with the transmission (e, f), calculated for a vapour with thickness $\ell = 1\ \mu$m and temperature T = 200 °C. The peak coupling Rabi frequency is

$\Omega_{15}^{\text{peak}} = 2\pi \times 2.5$ GHz, has a FWHM of 800 ps and is centered around $t = 0$. Before the pulse arrives, the steady-state solution maintains nearly all of the population in the ground states, distributed evenly between $|1\rangle$ and $|2\rangle$. As the pulse turns on, a large amount of population is excited into state $|5\rangle$, which oscillates throughout the duration of the pulse. The probe evolve differently, resulting in very distinct transmissive properties. With the probe on resonance, we predict a transient population inversion causing a gain feature to appear, whilst off resonance, the transmission oscillates around the steady-state value, with a transient increase in absorption. After the pulse, the system evolves gradually back to the steady-state solution. Again there is different behaviour between the on- and off-resonant cases. The signal from the resonant probe decays away relatively smoothly, whilst the off-resonant case oscillates due to the higher effective Rabi frequency of the probe ($\Omega_{\text{eff}} = \sqrt{\Omega^2 + \Delta^2}$).

9.2.2.1 Coupling Laser Power Dependence

We now examine the response of the system as the intensity of the coupling laser is varied. Figures 9.5 and 9.6 show the same two cases that were presented in Fig. 9.4 except that the peak Rabi frequencies are now also varied. The two plots are qualitatively very similar; as the Rabi frequency increases, the oscillations during the pulse become more rapid, allowing more total oscillations to take place within the same time window. The initial peak also moves towards the leading edge of the pulse, as expected. The main difference between the two, other than the scale of signal oscillation, is that off-resonance we predict the oscillations to include an increase in the absorption (orange/red on the colormap), whereas this increased absorption feature is not present when the probe is resonant.

9.2.2.2 Density Dependence

The density dependence of the Rabi oscillations is shown in Fig. 9.7. We include only the collisional broadening in the model, which shortens the excited state lifetime according to $1/\Gamma_{\text{tot}}$, where Γ_{tot} is given by Eq. (5.39). The main effect of the interactions is to broaden the oscillations which washes out the effect, and to push the peak of the first oscillation backwards in time. At the highest densities, the change in the probe transmission echoes the form of the pulse, centred around $t = 0$ with approximately the same FWHM.

9.3 Experiment

The experimental setup involves two lasers which copropagate through the nanocell. The beams have orthogonal linear polarisations, so that after the cell the beams can be separated with a polarising beam splitter. In addition to this polarisation filter

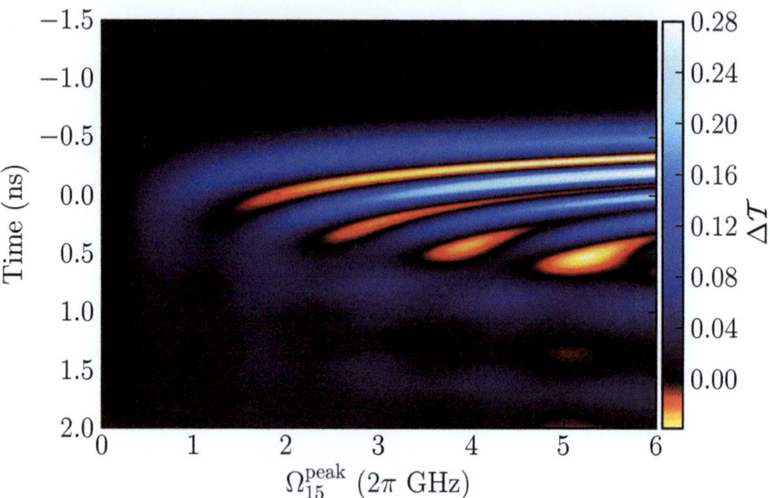

Fig. 9.5 Change in transmission $\Delta\mathcal{T}$ from the steady-state value as the peak coupling Rabi frequency is increased, with the probe off-resonance, $\Delta_p = 2\pi \times 1.5$ GHz. The other parameters are the same as for panels (a, c, e) of Fig. 9.4

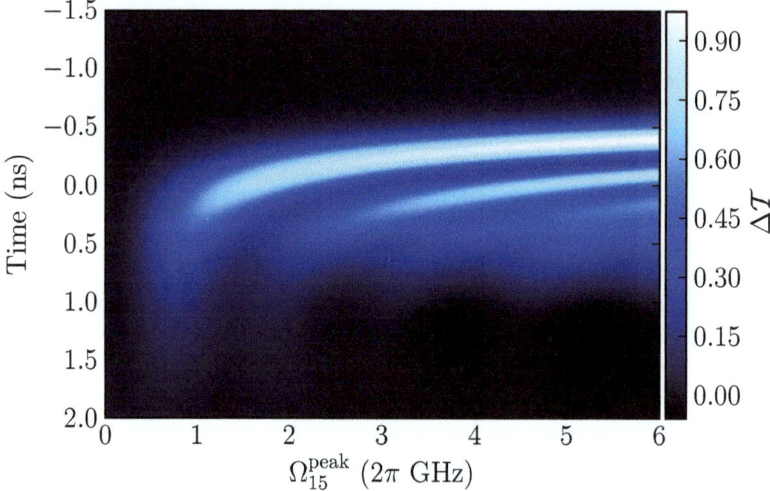

Fig. 9.6 Change in transmission $\Delta\mathcal{T}$ from the steady-state value as the peak coupling Rabi frequency is increased, with the probe on resonance, $\Delta_p = 0$. The other parameters are the same as for panels (b, d, f) of Fig. 9.4

(extinction $\sim$200), we use a Semrock FF01-800/12 bandpass filter, which passes 795 nm light with 95 % transmission, blocking 780 nm with a measured extinction of 3.5×10^3. With this setup, there is no detectable signal when the 780 nm pump

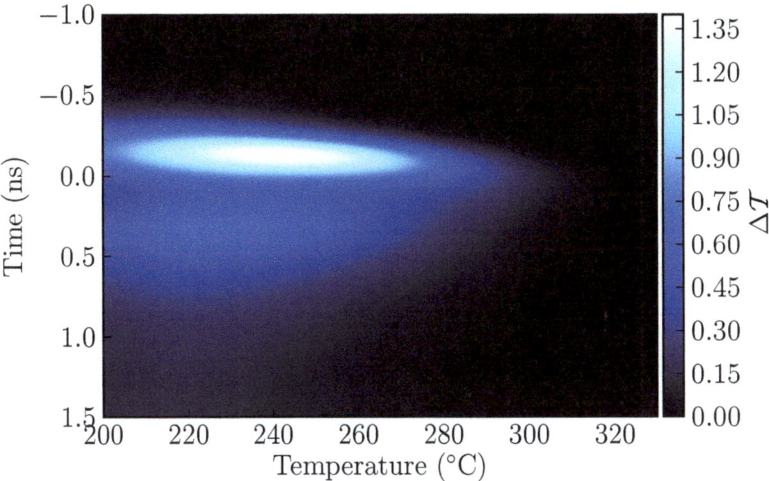

Fig. 9.7 Change in transmission $\Delta\mathcal{T}$ from the steady-state value as the temperature (density) is increased, with the probe resonant, $\Delta_p = 0$. The other parameters are the same as for panels (b, d, f) of Fig. 9.4

beam is on with the probe beam off. For detection we use a fast photodiode with 8 GHz bandwidth, which is the optical input to a PicoScope 9221A 12 GHz bandwidth sampling oscilloscope, with an effective sampling rate of $\sim$400 GS/s.

The probe beam is focussed to a waist of $\sim$10 μm whilst the coupling laser is focussed less tightly to a waist of $\sim$20 μm to minimise any intensity variation over the beam geometry, i.e. the atoms that see the probe beam all see roughly the same coupling beam intensity. The same optics are used for focussing but the initial size of the pump beam is smaller.

9.3.1 In-situ Rabi Frequency Measurement

Whilst the beam waists can be accurately measured without the cell present, the windows inevitably alter the focussing properties slightly. The only way to accurately measure the Rabi frequency that the atoms experience is by spectroscopic measurements. As in Fig. 9.3, we use the Autler-Townes splitting of the ground states to infer a Rabi frequency. The experimental data is shown in Fig. 9.8 as the (CW) coupling laser power is varied. The probe power for these measurements is $\sim$1 μW. The data looks qualitatively similar to the theoretical model in Fig. 9.3, but has some key differences. In an ideal three-level system, the splitting should be proportional to the Rabi frequency on the pump transition, which is proportional to the square-root of the laser power. For low powers we see this relationship, but as the power increases beyond $\sim$25 mW, the other states in the system become important. At this point

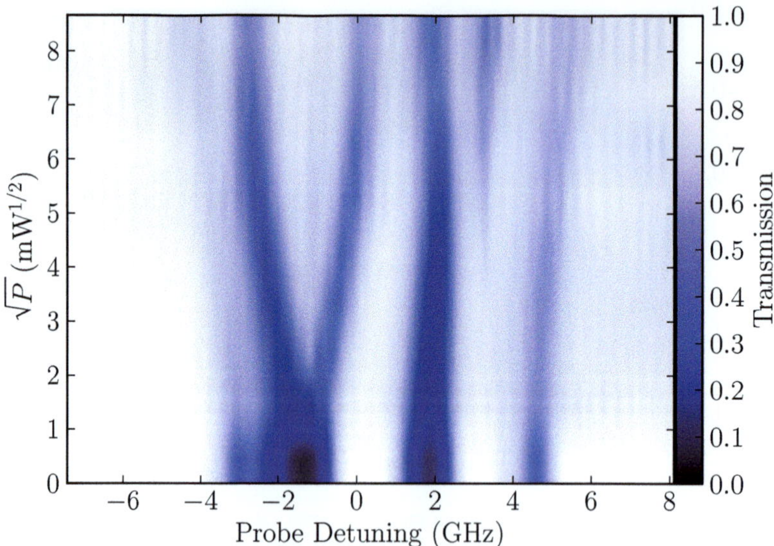

Fig. 9.8 Experimental data showing Autler-townes splitting, as a function of the power of the pump laser. The Rabi frequency is proportional to the square-root of the laser power, hence the splitting should vary linearly on this scale. For low powers we see this relationship, but as the power increases beyond ~25 mW, the splitting reaches around 3 GHz and does not increase further

the AT splitting reaches around 3 GHz and does not increase further. The exact mechanism for this apparent saturation is not well understood, but as the splitting at the point of saturation coincides with the ground state hyperfine splitting of ^{85}Rb, it is likely that the AT splitting of the lower ground state plays a role. In addition, the other isotope is clearly perturbed by the coupling beam, which is most clearly evidenced by the curving of the ^{87}Rb $F_g = 1 \rightarrow F_e$ transitions at an initial detuning $\Delta_p \sim 4.5$ GHz.

9.3.2 Rabi Oscillations

Using the measurement of the coupling Rabi frequency as a guide, and taking into account the relative transition strengths, we estimate the probe Rabi frequency to be of the order of 300 MHz for an optical power of 1 mW.

Figure 9.9 shows example signals of the variation in probe transmission when the medium is disturbed by the pulse, at a temperature $T = 215\ °C$. The pulse has a peak optical power of 85 mW. The blue curve is taken with a resonant probe laser, while the red data is the same conditions but with the probe detuned midway between the two hyperfine ground states of ^{85}Rb, $\Delta_p \approx 1.5$ GHz. The grey shaded area is the pulse intensity, measured on a fast photodiode and calibrated such that

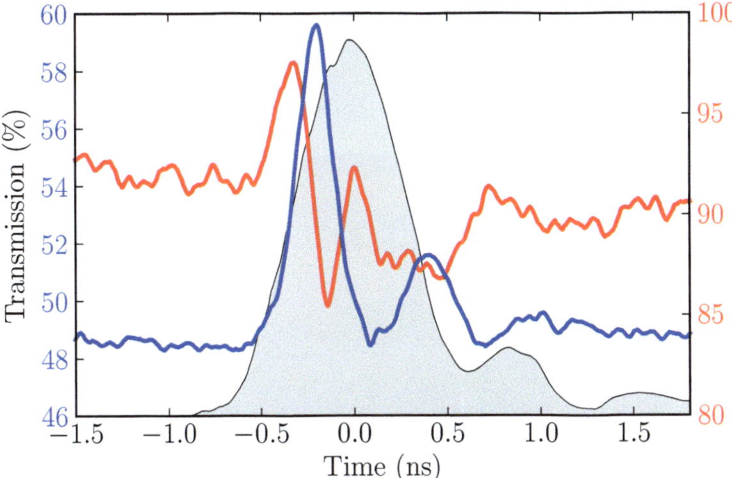

Fig. 9.9 Experimental data showing oscillatory behaviour at a temperature $T = 215\ °C$, vapour thickness $\ell = 2\ \mu\text{m}$, with (*blue*) $\Delta_\text{p} = 0$, and (*red*) $\Delta_\text{p} = 1.5$ GHz. The *red line* is the transmitted intensity of the CW probe, the *grey shaded area* is the measured pulse intensity (peak centred at $t = 0$)

when the pulse is detected on both detectors at the same time, the peaks overlap. As the PicoScope is a sampling (not real-time) oscilloscope, the data is necessarily an average over many pulse cycles. Systematic noise is removed from the signals by recording signals with the probe laser off. There is some qualitative agreement in the measured transmission profiles compared with the optical Bloch simulations of Fig. 9.4e, f. The transmission initially increases sharply with a rise time of around 200 ps, then oscillates with decreasing amplitude. In agreement with the simulation, the off-resonant probe experiences both enhanced absorption and transmission whilst the resonant probe sees only areas of enhanced transmission. However, for both sets of data the peak change in transmission is much smaller than the simulation.

The evolution of the experimental signals with varying coupling laser power is presented in Fig. 9.10. For this data the probe is detuned ($\Delta_\text{p} = 1.5$ GHz). Again, there is some qualitative agreement with the theory plot of Fig. 9.5. As the coupling laser power is increased the first peak of oscillations moves to earlier times, and becomes narrower. However, we do not see convincing evidence of multiple oscillations at the highest powers.

Figure 9.11 shows the change in transmission as a function of the vapour temperature, with coupling laser peak power of 85 mW and probe detuning $\Delta_\text{p} = 0$. As expected, the change in transmission grows as the temperature and hence optical depth is increased up to around 290 °C, after which the signal decreases again as the medium becomes too opaque. As the temperature is increased, the oscillations become less clear, which we attribute to the dephasing caused by the dipole-dipole interaction. At high density the temporal profile retains no oscillatory behaviour,

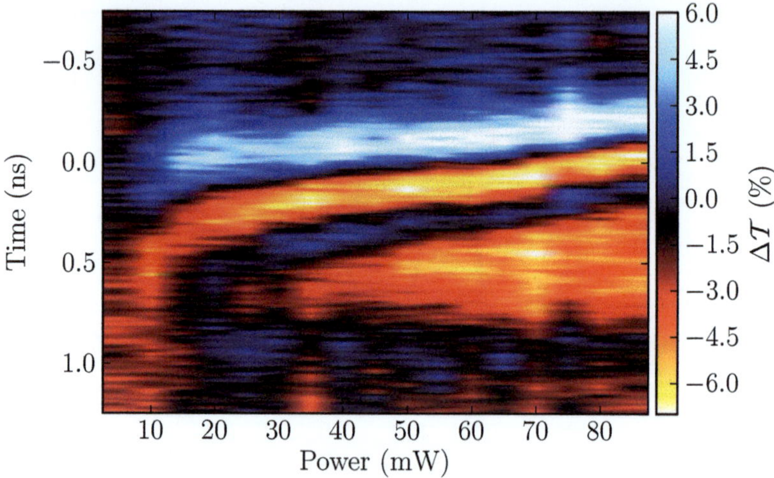

Fig. 9.10 Dependence of Rabi oscillations on the coupling laser power, at a temperature T = 215 °C, vapour thickness $\ell = 2\,\mu$m and probe detuning $\Delta_\mathrm{p} = 1.5$ GHz

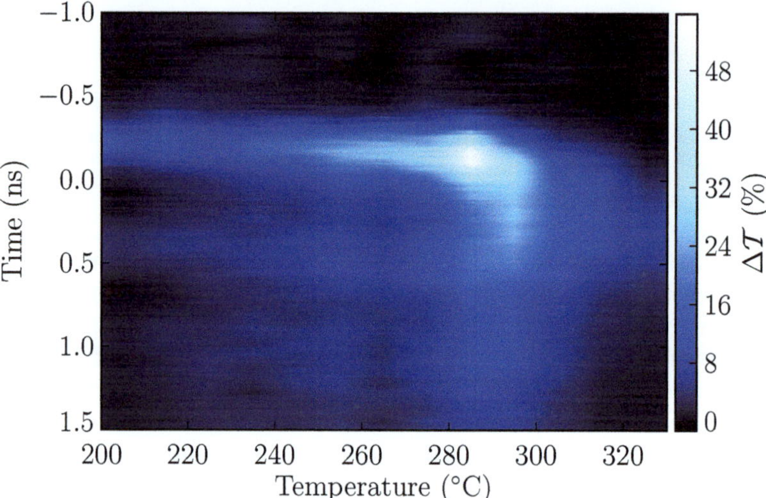

Fig. 9.11 Density dependence of Rabi oscillations, with vapour thickness $\ell = 2\,\mu$m, coupling laser power 85 mW, and probe detuning $\Delta_\mathrm{p} = 0$

instead roughly following the intensity envelope of the pump pulse. Whilst additional dephasing is expected at high density, there are two features of the data that are surprising. First, the peak of the data always occurs before the peak of the pump pulse. This is in contrast to the optical Bloch model which at high density roughly follows the pulse shape, peaked around $t = 0$. The second surprising element of the

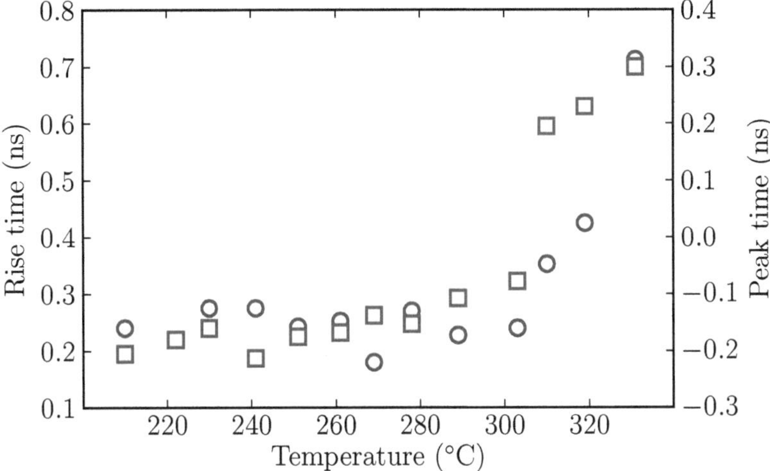

Fig. 9.12 Analysis of rise time (*blue circles*) and peak time (*red squares*) as a function of temperature, from the data presented in Fig. 9.11

data is that the rising edge of the peak remains sharp with a rise time of $\sim$200 ps. As a quantitative comparison, in Fig. 9.12, we plot a simple 10–90 % analysis of the rise time, in addition to the peak time. The two are correlated, and remain remarkably constant until the temperature rises above $\sim$310 °C (corresponding to a density $N = 3 \times 10^{16}$ cm^{-3}). The mechanism responsible for this behaviour is not well understood. It seems likely however that propagation effects, which are not part of the optical Bloch model, are important in the system. We can certainly assume that the medium is perturbed strongly by the pulse, so that the refractive index that the probe experiences changes dynamically as the pulse travels through the medium. Using the optical Bloch model, we can estimate how much the refractive index changes. This is plotted in Fig. 9.13, for the same parameters as panels (a, c, e) of Fig. 9.4. Given that the change in refractive index is significant, and the fundamental link between refractive index and pulse propagation (see Chap. 7), it is likely that propagation effects are important to the system. However, this level of simulation involves solving the full Maxwell-Bloch equations (see, for example, Ref. [6]) which is beyond the scope of this thesis.

9.4 Outlook

The observation of coherent dynamics in the nano-cell is ultimately limited by the wall-to-wall lifetime to a few nanoseconds. This forces the use of GHz Rabi frequencies and nanosecond excitation pulses. Whilst in this chapter we have shown some evidence of Rabi oscillations, the number of oscillations is limited. Ideally we would

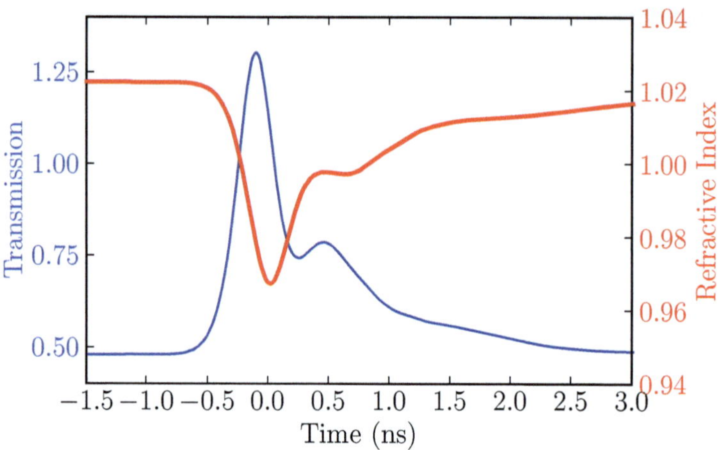

Fig. 9.13 Transmission profile and corresponding refractive index change calculated from the optical Bloch model. Parameters are the same for panels (a ,c, e) of Fig. 9.4

like to increase the number of oscillations, both to confirm beyond reasonable doubt of their origin and to study the dynamics further.

Apart from increasing the laser power which would require changing laser systems, we could focus more strongly to increase the intensity. Short focal length, high numerical aperture aspheric lenses will allow us to do this in the short term, whilst in the longer term building lenses into the cell window is also an option. These solid immersion lenses have been used successfully to study single atoms [7] and molecules [8] in recent experiments.

Alternatively, the pulse duration could be increased; pulse stretching techniques using either a ring or Fabry-Perot cavity are simple and inexpensive (termed a "poor man's pulse stretcher" [9]) optical techniques to do this and can yield around a factor of 1.5 increase in pulse FWHM, but with a factor of two loss in the peak intensity and some pulse shape distortion. A combination of increasing both the duration and focal intensity could provide similar peak Rabi frequencies but with longer duration, allowing the observation of more oscillations.

Extracting the underlying physics in the current experiment is difficult, as many of the experimental variables are connected in non-trivial ways. For example, by changing the density of the medium we not only influence the interactions between atoms, but change the optical depth of the medium as well. Decoupling these processes would be beneficial. Therefore in the longer term, a more significant change in the experiment may prove advantageous. Whilst the Vee-type energy level arrangement in the current implementation optimises the Rabi frequency for a given laser power because of the strength of the atomic transitions, moving to a ladder system similar to that used in Ref. [10] should give cleaner signals. A combination of a pump pulse on the ground state transition to the $5P_{3/2}$ level with a CW probe on a transition from $5P_{3/2}$ to a higher excited state would also yield a zero-background signal.

References

1. T. Baluktsian, B. Huber, R. Löw, T. Pfau, Evidence for strong van der Waals type Rydberg-Rydberg interaction in a thermal vapor. Phys. Rev. Lett. **110**, 123001 (2013)
2. P. Siddons, C.S. Adams, C. Ge, I.G. Hughes, Absolute absorption on rubidium D lines: comparison between theory and experiment. J. Phys. B **41**, 155004 (2008)
3. D.A. Steck, Rubidium 85 D Line Data (2009)
4. M. McDonnell, D. Stacey, A. Steane, Laser linewidth effects in quantum state discrimination by electromagnetically induced transparency. Phys. Rev. A **70**, 1 (2004)
5. D.N. Stacey, D.M. Lucas, D.T.C. Allcock, D.J. Szwer, S.C. Webster, Optical Bloch equations with multiply connected states. J. Phys. B **41**, 085502 (2008)
6. P. Siddons, Faraday Rotation of Pulsed and Continuous-wave Light in Atomic Vapour, PhD thesis, Durham University, 2011
7. S.A. Aljunid et al., Phase shift of a weak coherent beam induced by a single atom. Phys. Rev. Lett. **103**, 153601 (2009)
8. M. Pototschnig et al., Controlling the phase of a light beam with a single molecule. Phys. Rev. Lett. **107**, 063001 (2011)
9. W. Martin, Pulse stretching and spectroscopy of subnanosecond optical pulses using a Fabry-Perot interferometer. Opt. Commun. **21**, 8 (1977)
10. B. Huber et al., GHz Rabi flopping to Rydberg states in hot atomic vapor cells. Phys. Rev. Lett. **107**, 243001 (2011)

Chapter 10
Project Outlook

Many fundamental questions still remain unanswered about the interaction between atoms and light, particularly when strong interactions between the atoms are present. For optical wavelengths, studying these interactions in ultracold systems is very difficult, as the densities required are prohibitively large. Using thermal atoms to study the same physics is therefore attractive because realising a high density atomic vapour is relatively easy. We have shown in this thesis how studying these strong interactions is plausible. With strong enough interactions, the motional effects that are the inescapable consequence of using thermal vapours can become relatively unimportant. In this ultra-high density regime, we once again recover the coherent character of the dipole-dipole interaction between atoms, as evidenced in the observation of the cooperative Lamb shift presented in Chap. 5.

However, the observed interaction is still an angular average. We envisage optimising the dipole–dipole interaction by controlling the positions of the dipoles in the vapour. A new generation of vapour cells constructed with micro-channels similar to those in Ref. [1] could enable this control by providing an additional degree of spatial confinement.

A recent suggestion by Kaiser [2] that dense atomic vapours could be a good system to look for Anderson localization of light [3, 4]. Whilst much effort is focussed on increasing the density in ultracold atomic systems [2], the observation of coherent interactions in thermal ensembles questions the widely held view that collisional or motional processes will destroy the coherence. A high density thermal atomic vapour could therefore be a promising system to look for localisation effects.

Motional effects can also be partially suppressed by moving to pulsed excitation, and with strong enough optical pulses it is possible to observe Rabi oscillations on a very short timescale, much faster than the dephasing mechanisms in the system. As of the time of writing this thesis, relatively little is understood about the behaviour of this strongly driven system, and it remains an active area of study.

J. Keaveney, *Collective Atom–Light Interactions in Dense Atomic Vapours*, 127
Springer Theses, DOI: 10.1007/978-3-319-07100-8_10,
© Springer International Publishing Switzerland 2014

Research into cooperative processes is a vibrant and fruitful area, and as groups revisit old problems with new technology the boundaries are constantly being pushed. Thermal atom experiments are enjoying a renaissance at the moment. Long may it continue.

References

1. T. Baluktsian et al., Fabrication method for microscopic vapor cells for alkali atoms. Opt. Lett. **35**, 1950 (2010)
2. R. Kaiser, Quantum multiple scattering. J. Mod. Opt. **56**, 2082 (2009)
3. P.W. Anderson, Absence of diffusion in certain random lattices. Phys. Rev. **109**, 1492 (1958)
4. M. Segev, Y. Silberberg, D.N. Christodoulides, Anderson localization of light. Nat. Photonics **7**, 197 (2013)

Appendix A
Nano-Cells

This appendix details the properties of the thin cells used in this work, and in particular how the length is calibrated in the nano-cell.

A.1 The Nano-Cell

All of the vapour cells (except for reference cells) used in this thesis were constructed in the group of David Sarkisyan, in the Institute for Physical Research, Yerevan, Armenia. They are the source of the technical details in this Appendix.

The cells are constructed from a Rb reservoir (often called the 'side-arm' in publications) attached to the cell windows. The reservoir is relatively far away from the windows, allowing for a temperature gradient between the two which prevents condensation of Rb on the windows. During the experiments performed in this thesis, the windows are kept roughly 50 °C hotter than the reservoir. Experimentally, we find the reservoir temperature determines the vapour pressure.

A schematic of the nano-cell is shown below in Fig. A.1, see Chap. 1 for a photo of the window regions and additional description. The cell windows have a wedge-shaped profile, allowing vapour thickness to be tuned between 30 nm and $2\,\mu$m. These windows are constructed from two sapphire (refractive index $n = 1.8$) plates, with the c-axis aligned with the propagation axis to eliminate any birefringence. One of these plates is slightly curved, with a radius of curvature $R > 100$ m. This leads to the Newton's Rings interference pattern visible in the photo. The region of minimum thickness is at the centre of these rings. The sapphire plates are of exceptional quality, with a local surface roughness of <3 nm over any 1 mm^2 region. Though there may be more variation over the entire surface, this is unimportant as the laser focus is always much smaller than 1 mm^2.

J. Keaveney, *Collective Atom–Light Interactions in Dense Atomic Vapours*,
Springer Theses, DOI: 10.1007/978-3-319-07100-8,
© Springer International Publishing Switzerland 2014

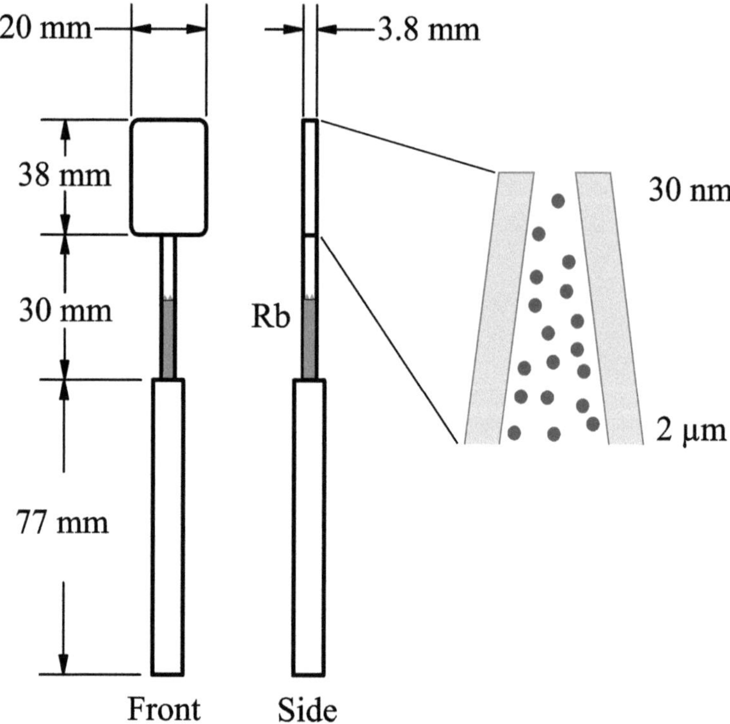

Fig. A.1 Nano-cell schematic. The cell has a wedge-shaped profile, allowing for a tunable vapour thickness of between 30 nm and 2 μm, constructed from two extremely flat sapphire plates. One of these plates is slightly curved, leading to the Newton's Rings interference pattern (see Chap. 1). The Rb reservoir is relatively far away from the windows, allowing for a temperature gradient between the two which prevents condensation of Rb on the windows

A.1.1 Operation of the Thin Cells

During the operation of these cells, care should be taken to heat and cool gradually, to ensure the glass does not fracture due to a large temperature gradient. There is no hard limit, but in this author's experience, 10 °C/min is a rough guide to a maximum heating/cooling rate. For heating, the windows should be heated first, to establish a temperature difference with the reservoir; this prevents condensation of Rb on to the windows.

Should condensation occur, heat the windows only, to around 300 °C and leave for a couple of hours. The temperature gradient should be enough to promote the Rb off the windows and back into vapour. After this time, cool very slowly over a further couple of hours back to room temperature.

After the windows have heated, heat the reservoir to the desired temperature. Thermal expansion has been taken into account and is linear with temperature with a gradient of approximately 1 nm for every 5 °C of window temperature.

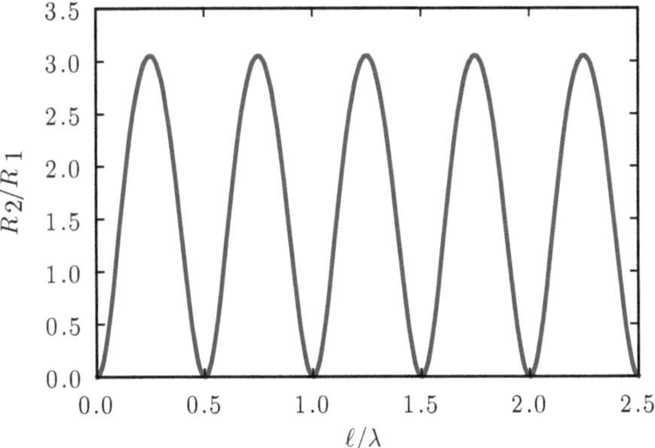

Fig. A.2 $\cos^2$ oscillations of the ratio of reflections R_2/R_1, which is used for thickness calibration

A.2 Thickness Calibration

The local thickness in the nano-cell can be determined using an interferometric method which utilises the Fabry-Perot nature of the cell. The implications of this were detailed in Chap. 3. Figure 3.6 shows the reflections from the surfaces in the nano-cell. R_1 is from the first window of the cell, at the air-sapphire interface, and therefore this is constant and used as a reference. R_2 is from the sapphire-Rb interface, and as detailed in Chap. 6, undergoes periodic oscillations with the thickness of the medium. If we assume that the refractive index of the Rb layer is approximately unity, valid for far off-resonant light, then the ratio of the reflections R_2/R_1 can be described analytically using a standard Fabry-Perot treatment as [1]

$$\frac{R_2}{R_1} = \frac{(1 - R^2)F \sin^2 \phi/2}{R(1 + F \sin^2 \phi/2)}, \tag{A.1}$$

where

$$F = \frac{4R}{1 - R^2}, \tag{A.2}$$

$$R = \left(\frac{n-1}{n+1}\right)^2, \tag{A.3}$$

$$\phi = 4\pi\ell/\lambda, \tag{A.4}$$

and n is the refractive index of the glass. For sapphire ($n = 1.8$) windows, this leads to the thickness variation shown in Fig. A.2, which oscillates with a period of $\lambda/2$.

The ratio is of course independent of the laser power and insensitive to any fluctuations, so can be determined accurately. This measurement is still not conclusive, however, so the exact thickness must be determined from examining the spectral data—for example, where the strongest Dicke Narrowing features coincide with $R_2/R_1 \rightarrow 0$, then the thickness is $\lambda/2$. One can also determine the gradient of the slope by moving the cell to thicker/thinner regions and observing how R_2/R_1 changes.

A.2.1 Finite Beam Size

Due to the finite beam size and the geometry of the cell, there will be a small variation in the thickness of the cell. We can estimate how much of a problem this is with a simple geometric argument. If we assume the wedge profile forms a triangle, with perpendicular sides of 30 mm (height) and 2 μm (width), we can use similar triangles to work out how much the thickness varies over a given beam size

$$\Delta \ell \approx 2w_0 \times \frac{2 \ \mu\text{m}}{30 \ \text{mm}}, \tag{A.5}$$

where $2w_0$ is the $1/e^2$ beam diameter. For a typical experimental waist $2w_0 = 20 \ \mu$m, this gives $\Delta \ell < 2$ nm.

Appendix B
Experimental Details for Transmission Spectroscopy

This appendix details the methods used for properly calibrating experimental transmission spectra. There are two axes to properly calibrate—the frequency (detuning) axis from the laser scan, and the transmission axis from the laser intensity fluctuations.

B.1 Experimental Setup

A typical experimental setup is shown below in Fig. B.1. We use pump-probe Doppler-free (saturated absorption/hyperfine pumping [2]) spectroscopy in a reference cell to provide an absolute frequency reference, and a Fabry-Perot Interferometer (FPI) to correct for non-linearities in the laser scan. Both of these signals are recorded simultaneously with the experimental spectra in the thin cell. For the thin cell spectra, the light is linearly polarised and focussed to a small spot size (typically we have a $1/e^2$ radius of 10 μm using lenses with focal length $f = 100$ mm) to minimise thickness fluctuations due to the cell geometry (see Appendix A). Even so, the uncertainty in cell thickness is limited primarily by the thickness fluctuations to around 5 nm, depending on the focal spot size. The optical power is attenuated using neutral density filters, typically to less than 1 μW, and we record the measured spectra on conventional photodiodes.

B.2 Data Processing and Normalisation

The entire processing, normalisation and fitting routines have been streamlined into a single, interactive program written in Python with a GUI written in wxPython.

J. Keaveney, *Collective Atom–Light Interactions in Dense Atomic Vapours*,
Springer Theses, DOI: 10.1007/978-3-319-07100-8,
© Springer International Publishing Switzerland 2014

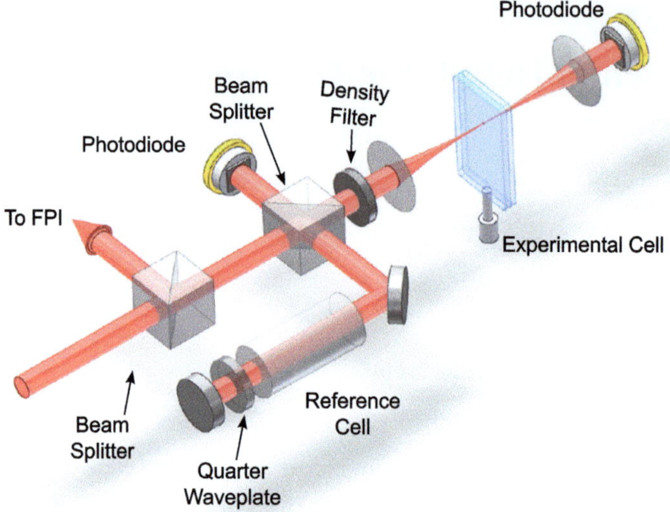

Fig. B.1 Experimental setup for transmission spectroscopy. A 7.5 cm reference vapour cell is used in a pump-probe sub-Doppler spectroscopy setup for frequency reference, and a Fabry-Perot Interferometer (FPI) for correcting non-linearity in the frequency scan of the laser. Typically $f = 100$ mm focal length lenses are used to focus in to a specific thickness (with uncertainty in the vapour thickness typically <5 nm). Neutral density filters before the cell ensures that optical pumping is suppressed

B.2.1 Calibrating Frequency Axis

Calibration of the frequency axis is done in two parts; linearisation of the time axis, and scaling. The process employed here is similar to that in Ref. [3]. We require an accurate time-to-frequency calibration, which is done using the cavity resonances from an FPI, which occur at equally spaced frequencies. A typical FPI signal is shown in Fig. B.2(a). An analysis routine was written to detect the peak positions (red vertical lines). Only data points with signal values above a threshold level are retained. In this array, the peaks are separated by analysing the time coordinates; if there is a discontinuous jump in the time coordinate, this is recognised as another discrete peak and the array is thus subdivided into smaller arrays centered around each of the peak values. The exact peak positions are found from the maxima of these sub-arrays.

The peak positions are fitted to a straight line, which represents linear time to frequency conversion. However, there is a characteristic deviation from this, because the laser scan has a small non-linearity. This is evident if we plot the deviation from the linear fit, as shown in panel (b). The deviation is fitted to a 5th order polynomial function[1] and the time axis is adjusted accordingly to remove this non-linearity.

---

[1] Higher orders can be used but give diminishing returns. 5th order represents a good tradeoff between fitting accuracy and computational speed.

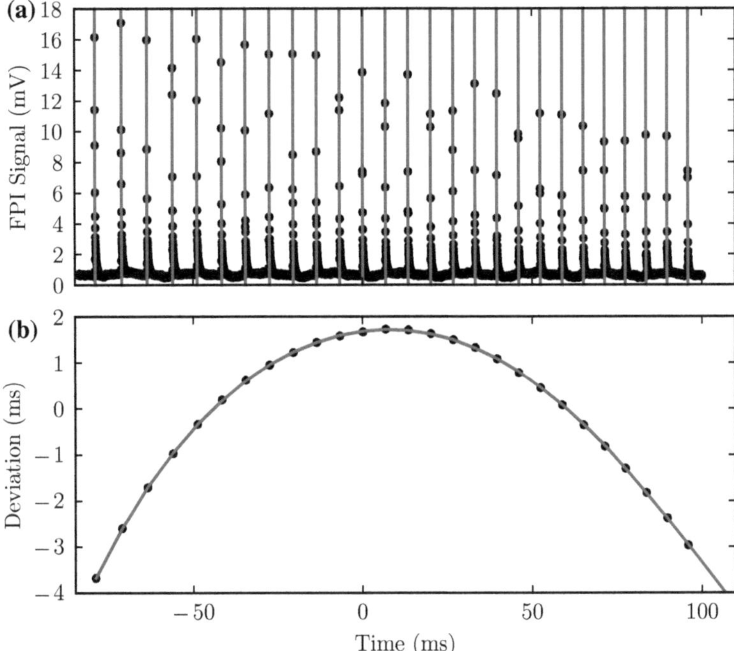

Fig. B.2 **a** The Fabry-Perot signal (*black markers*, not all data points shown) and detected peak positions (*red*). **b** The deviation of the peak positions from a linear fit. The removal of this deviation linearises the time-to-frequency scaling

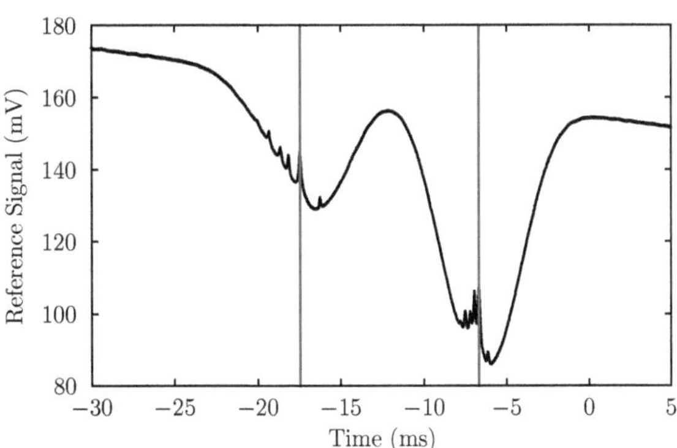

Fig. B.3 Absolute frequency calibration using sub-Doppler features. This example is from the D2 line, using the crossover resonances detailed in the main text

The next step is to calibrate the frequency axis, which is done using two peaks from the pump-probe spectrum. There are several features in each Doppler-broadened absorption line (six for the D2 line, 3 for the D1 line) corresponding to the excited state hyperfine levels and the crossover resonances [2]. An example of this is shown in Fig. B.3. As the crossover resonances are the stronger, we use these to calibrate the frequency axis. Since the time-to-frequency conversion is now linear, we only require

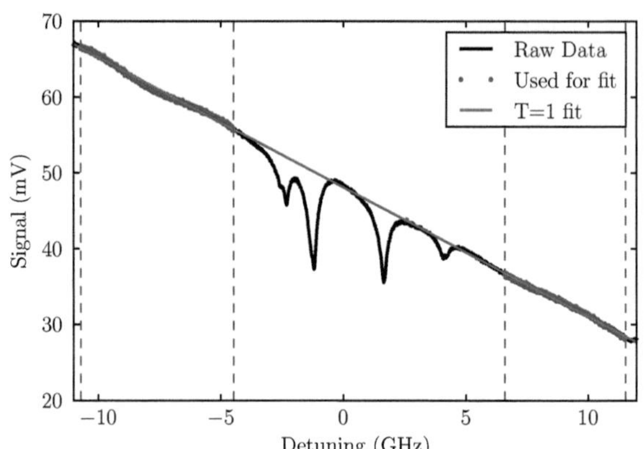

Fig. B.4 Normalisation of transmission axis using the off-resonant areas of the transmisison spectrum. The areas of the spectrum highlighted in blue are fitted to a polynomial function to determine the transmission scan, which is shown in red. The normalised transmission is then the raw data divided by this curve

two points to fully calibrate the frequency axis. Any points of known frequency will do but in this work, for the D2 line we use the $F_g = 2 \rightarrow F_e = 2, 3$ and $F_g = 3 \rightarrow F_e = 3, 4$ crossover resonances in ^{87}Rb and ^{85}Rb, respectively, as these are the strongest features. For the D1 line, we use the $F_g = 2 \rightarrow F_e = 1, 2$ and $F_g = 3 \rightarrow F_e = 2, 3$ crossover resonances in ^{87}Rb and ^{85}Rb, respectively.

B.2.2 Calibrating Transmission Axis

Due to the feed-forward in the laser electronics, as the voltage to the piezo controlling the fine position of the grating is scanned, the current is also modified, with the aim of increasing the mode-hop-free scan range. The result is an asymmetric signal on the photodiode, as the output power changes with the frequency of the laser, which needs to be corrected.

To correct for this, a spectrum is recorded when the cell is relatively cold and there is no absorption in the wings of the absorption lines. As the temperature (and number density) increases, the absorption lines become broader such that the spectrum is wider than the mode-hop-free tuning range of the laser, so in this case we make the assumption that the scan profile does not change over the time it takes to run the experiment.

With the low temperature scan, we can then select areas of the spectrum that are known to have no absorption. This is implemented interactively, so the user selects the regions of the spectrum graphically, significantly reducing post-processing time for any given data set.

After this, the selected areas are fitted to a polynomial function (usually 2nd order is enough) and this function is used as the $\mathcal{T} = 1$ line, the data is divided by this function to compute the normalised transmission curve. An example plot showing this process is shown in Fig. B.4.

Appendix C
SPCM Data Acquisition

This appendix details the data acquisition method for photon counting, used in Chaps. 7 and 8. Using the statistics methods built in to a LeCroy WaveRunner 6Zi oscilloscope, it is possible to implement photon counting without the use of any complex counting electronics. The advantages are speed of data acquisition, ease of setup, immediate real-time signals and the ability to use photon counting in tandem with signals from photodiodes etc.

C.1 Setup

The SPCM used in this work is a Perkin Elmer SPCM-AQR-14-FC, which has an optical fibre input and a BNC output. This makes it very easy to integrate into a setup. This is connected *directly* into the oscilloscope; no other electronics are needed. Optionally, the SPCM has a gate input which allows the user to turn off the device, otherwise it is open constantly. For the work in this thesis, the SPCM was operated without gating.

The SPCM 'clicks' and sends an electrical pulse on a photon detection event. The first thing to do is to look at the pulse shape of a single click. To do this, set the scope to trigger on the same channel as the SPCM is on. Figure C.1 shows the detected pulse shape. The pulse window is around 15 ns, therefore in order to detect every electrical pulse from the SPCM, the oscilloscope requires a sampling rate of at least $1/(15 \times 10^{-9}) = 66$ MS/s. For experiments where the time bins of the histogram are wider than 15 ns, all that is required is to detect a single peak position, and build up the histogram from this. However, for more time sensitive measurements then we use a more complex acquisition method. The most repeatable way of detecting the pulse is not to find the maximum value of the electrical signal, as in panel (a), but to find the maximum of the (highly smoothed) time derivative, which is shown in panel (b). As the turn-on of the pulse is spread over $\sim$2 ns, the time resolution is not limited by the bandwidth of the oscilloscope, but by its sampling rate—at a maximum sampling rate of 20 GS/s, this gives a 50 ps time resolution.

J. Keaveney, *Collective Atom–Light Interactions in Dense Atomic Vapours*, Springer Theses, DOI: 10.1007/978-3-319-07100-8,
© Springer International Publishing Switzerland 2014

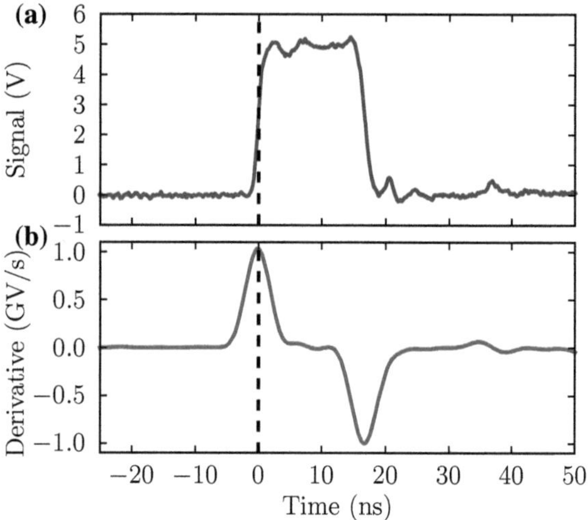

Fig. C.1 a Electrical pulse shape of a single 'click' of the SPCM. **b** Smoothed time derivative of (**a**), the peak of which is used as the measure of photon arrival time

The aim is to time the arrival of photons, so a reference is needed—in practice this is whatever triggers the oscilloscope; for our purposes we trigger from the Pockels cell driving electronics (Chaps. 7 and 8) or from the piezo trigger output of the laser electronics.

Whilst in principle every acquisition, and hence the time information of every photon arrival, can be saved directly, the amount of storage required quickly becomes prohibitive without additional processing. For example, a single acquisition of the data presented in Fig. C.3 is 250 MB, owing to the large sampling rate (100 MS/s) and time window (200 ms). For the pulsed light experiments presented in the main body of this thesis, the file sizes for each acquisition are smaller, around 100 kB, but as the pulse repetition rate is around 1 kHz, this gives a data generation rate of around 100 MB/s, assuming the hard disk can keep up with this rate. Assuming a hard disk capacity of 500 GB, this would be completely filled in around an hour. Typical experiments take around 2 hours to detect enough photons for a good signal-to-noise ratio, so clearly a better method is needed. If detailed time information of every acquisition is required, it is possible to interface with the oscilloscope via a VisualBasic Script (*.vbs file), for real-time data processing.

To record our data, we take advantage of a software feature on the oscilloscope, which generates in real-time a histogram of a specified parameter, in this case the photon arrival time, with up to 5,000 time bins. For CW fluorescence data, this gives us an approximate frequency resolution of 2 MHz, assuming the laser scans over 10 GHz in the 200 ms interval, though this is only limited in principle by the scan range. For the pulsed experiments, we typically have a time window of 10 ns (fast light) or a few hundred ns (excited state decay), allowing for time bins spaced by picoseconds, if desired.

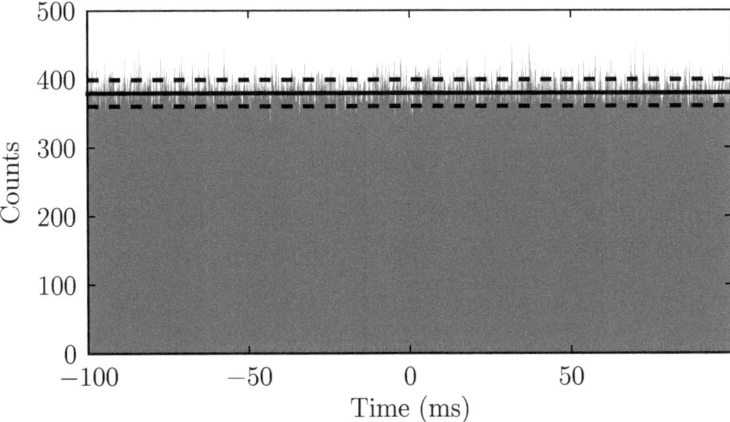

Fig. C.2 Histogram of the dark counts measured on the SPCM. The *solid line* is the mean count number, $\mu = 380$, while the *dashed lines* are $\mu \pm \sigma$, where $\sigma = 20$ is the standard deviation. The measured value of σ is very close to that expected from Poissonian statistics, where we would expect $\sigma = \sqrt{\mu} = 19.5$

C.2 Testing

C.2.1 Detector Dark Counts

The dark count is a feature of all counting electronics, and is the signal generated in the absence of any external stimulus. This signal should be uniformly distributed in time, at a specified rate. For our detector, the manufacturer specified rate is 50 Hz. Figure C.2 shows the measured distribution of dark counts on the detector, which we measure with the detector turned on but with no light present. As expected the distribution is flat over time. The total detection time $t_{\mathrm{d}} = t_{\mathrm{acq}} n_{\mathrm{acq}}$ is the interval per acquisition multiplied by the number of acquisitions. The count rate $\dot{C}$ is then number of counts, N, divided by t_{d}. For the data in Fig. C.2, at room temperature $n_{\mathrm{acq}} \approx 75 \times 10^3$, $t_{\mathrm{acq}} = 200$ ms, and $N \approx 750 \times 10^3$, yielding $\dot{C} = N/t_{\mathrm{d}} = 49$ Hz.

C.2.2 Saturation Effects

The SPCM is very sensitive to the amount of incident light, with a non-linearity in the count rate that is well documented in the Perkin Elmer datasheets. One aspect of this non-linearity that is not well discussed is the temporal response. We have found that increasing the incident light level causes the histogram of detected counts to be distorted, moving the detected signal backwards in time. Care should therefore be taken to ensure that for time-sensitive measurements—such as the optical pulse

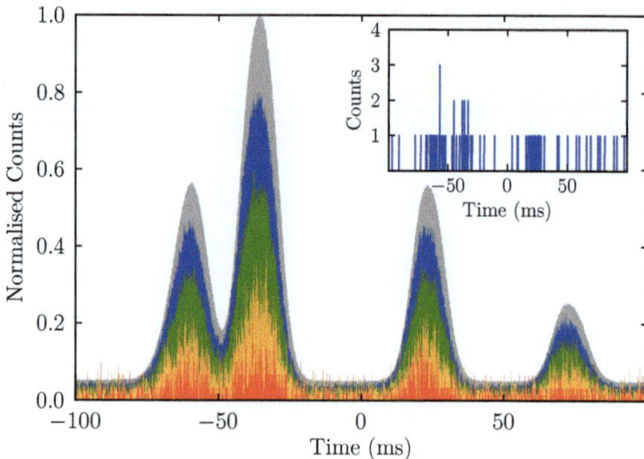

Fig. C.3 Fluorescence as a function of laser detuning (time, uncalibrated) across the D2 resonance, for total count numbers 10^3 (*red*), 10^4 (*yellow*), 10^5 (*green*), 5×10^5 (*blue*), and 25×10^6 (*grey*); for clarity the peak counts have been normalised to 0.2, 0.4, 0.6, 0.8 and 1, respectively. Signal-to-noise improves with total counts, as expected. For comparison, the inset shows the data acquired from a single trigger, where we detect around 100 photons

propagation experiment in Chap. 7 of this thesis—the count rate (or CW equivalent count rate for pulsed excitation) is kept well below the maximum level.

In addition, using too much power also causes the detector to be constantly saturated. The histogram in this case is made up of many peaks separated roughly by the dead-time of the detector.

To be clear—for experiments where time sensitive detection is required, the light level must be kept low enough so that the detector is not saturated. As a rule-of-thumb, in our experiments we keep the below 10 % of the maximum detector count rate.

C.3 Applications

In addition to the applications presented in the main body of this thesis (Chaps. 7 and 8) in this section we present two other methods that may prove to be useful in future experiments.

C.3.1 Frequency-Dependent Fluorescence

It is possible to acquire photon arrival time data using any time reference as a trigger. In Chaps. 7 and 8, we triggered on a reference from the Pockels cell that generates the optical pulses. However, it is also possible to use the SPCM in a way similar to a conventional photodiode, with a CW light source, and detect fluorescence.

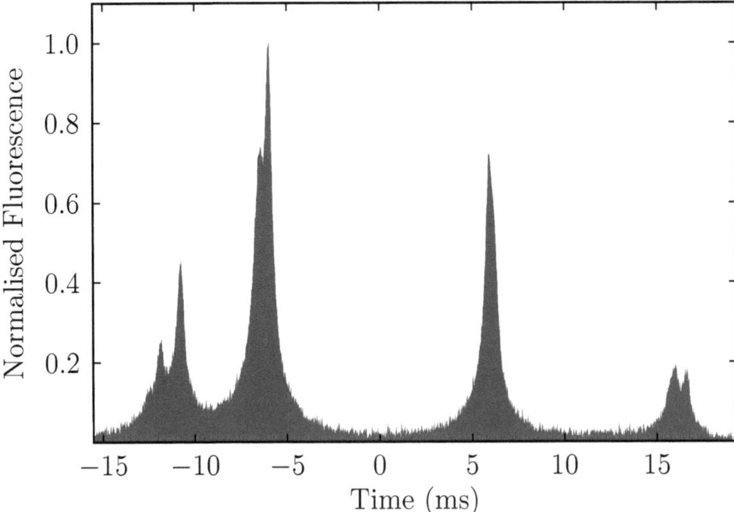

Fig. C.4 Example fluorescence spectrum with a vapour thickness $\ell = \lambda$, at a temperature $T = 100\,°\mathrm{C}$, clearly showing Dicke narrowing with good signal-to-noise and excellent resolution of the hyperfine features. For comparison, we would expect to absorb a maximum of 0.5 % of the incident light under these conditions, making transmission spectroscopy difficult

Figure C.3 shows the fluorescence signal obtained as a function of the laser detuning over the D2 resonance line in Rb, where the time reference (trigger) is linked to the laser frequency scan. This figure also shows how the signal-to-noise evolves as the number of counts increases; for a single shot we might obtain around 100 individual detection events, as shown in the inset (this number can be considerably higher, a low count rate is shown only as an example). The signal-to-noise ratio improves dramatically with the total number of detection events, as expected from Poissonian counting statistics.

With the thin cell, this can be a particularly useful technique, as the SPCM is much more sensitive than a conventional photodiode. This allows investigation of vapours where the optical depth is very low, and hence the fluorescence intensity is also very small. In Fig. C.4, we show example data of fluorescence detected in a cell of thickness $\ell = \lambda$ at a temperature $T = 100\,°\mathrm{C}$ as the laser is scanned over the Rb D2 line. In a transmission spectroscopy experiment, we would expect to absorb only around 0.5 % of the incident light, yielding a very small signal. The hyperfine structure is well resolved because of Dicke narrowing (see Chap. 3), and even though the absorption is very low the signal-to-noise is good enough to see these features clearly. CW fluorescence measurements therefore have potential for investigating media with very low optical depth.

C.3.2 Application: Correlation Measurement, $g^{(2)}(\tau)$

This needs two SPCM modules. One detector should be used as the trigger, then make a histogram of the photon arrival time of the other detector. When the first detects a photon, it triggers the scope, and a point on the histogram is generated if the second detector also observed a photon in the same time window. The time difference between the two detectors can then be measured.

An easy trial of this is to connect the two inputs of the detectors with the same optical fibre. The dark count of one detector sometimes generates a photon which will travel down the fibre and be detected by the other detector. The correlation histogram should then have a trough at $t = 0$ with peaks at $t_{\pm} = \pm L/c_f$, where L is the length of optical fibre and c_f is the speed of light in the fibre. All other times should be uniformally distributed due to the dark counts and any background light entering through the cladding of the fibre.

Appendix D
Large Susceptibility

Since much of this work deals with large susceptibilities, the usual approximation of $n \approx 1 + \chi/2$ does not hold, and one must use the full form $n = \sqrt{1 + \chi(\Delta)}$.

However, this is only valid provided one compensates for local field effects, which were discussed in Chap. 5. If one does not consider local field effects, $\sqrt{1 + \chi(\Delta)}$ is also a bad approximation, as it leads to an asymmetry in both the real and imaginary parts of the susceptibility. Figure D.1 shows this asymmetry as the magnitude of χ is increased. The magnitudes of the plotted refractive index profiles are experimentally feasible (see Chap. 6).

We can estimate where the approximation breaks down by looking at the value of the susceptibility when local field effects become important. On resonance, the imaginary part of the susceptibility is a maximum, and is given by

$$\mathrm{Im}\chi = \frac{N\mu^2}{\hbar\epsilon_0} \frac{2}{\Gamma_0 + \beta N}. \tag{D.1}$$

We assume that local field effects become important for an atomic density such that the cooperativity parameter $\mathcal{C} = 1$. Using the relationship between dipole moment and natural linewidth (for a $J = 1/2$ to $J = 1/2$ transition), Eq. (5.9), including self-broadening and assuming $N = 1/r^3$, Eq. (D.1) reduces to

$$\mathrm{Im}\chi = 0.5 \tag{D.2}$$

In a real system, hyperfine structure reduces this value further. Assuming the populations are evenly distributed throughout the Zeeman sublevels of the ground state, for the strongest hyperfine transition on the D2 line, the maximum value is around 0.07, equivalent to the blue curve in the figure. Assuming the only shift is due to the Lorentz shift, for this density there is a redshift of $\Delta_{LL} = -N\mu^2/3\hbar\epsilon_0 = -\Gamma_0$, much larger than the apparent shift due to the asymmetry.

J. Keaveney, *Collective Atom–Light Interactions in Dense Atomic Vapours,*   143
Springer Theses, DOI: 10.1007/978-3-319-07100-8,
© Springer International Publishing Switzerland 2014

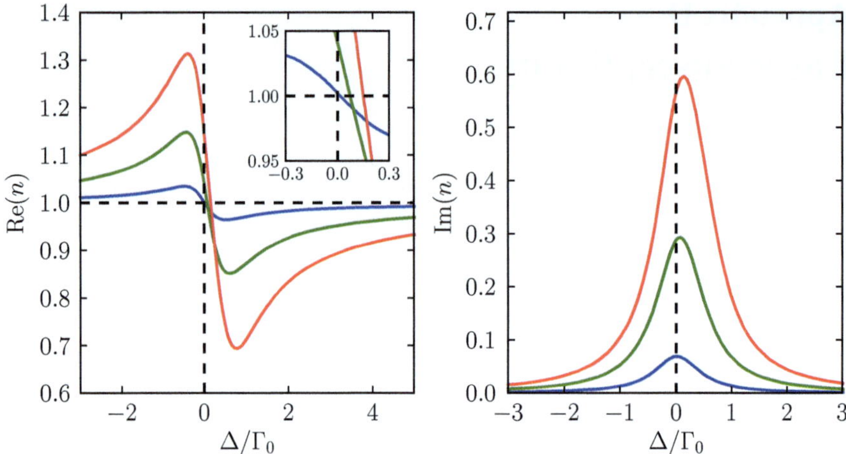

Fig. D.1 The real and imaginary parts of the refractive index $n = \sqrt{1 + \chi}$ are plotted for χ, without taking into account local field effects. At large values of χ, the lineshape is asymmetric and appears *blue-shifted* relative to the resonance frequency. In practice, however, the underlying physics changes as χ becomes larger so that the local field must also be considered, which introduces a *red-shift*, as discussed in Chap. 5

References

1. W.T. Silfvast, *Laser Fundamentals*, 2nd edn. (Cambridge University Press, Cambridge, 2004)
2. D.A. Smith, I.G. Hughes, The role of hyperfine pumping in multilevel systems exhibiting saturated absorption. Am. J. Phys. **72**, 631 (2004)
3. P. Siddons, C.S. Adams, C. Ge, I.G. Hughes, Absolute absorption on rubidium D lines: comparison between theory and experiment. J. Phys. B **41**, 155004 (2008)